```
Hill, Cherry,
   1947-

Cherry Hill's horse
care for kids.
```

$23.95

DATE			

Cherry Hill's
HORSE CARE for KIDS

Cherry Hill's

HORSE CARE
for KIDS

The mission of Storey Publishing is to serve our customers by publishing practical information that encourages personal independence in harmony with the environment.

Edited by Deborah Burns, Anne Kostick, and Eileen Clawson
Cover and text design by Wendy Palitz
Art direction by Meredith Maker
Text production by Susan Bernier and Leslie Tane

Front cover photographs: © (clockwise from top left) Shelley Heatley, Lesley Ward, © Benson Photography, Shelley Heatley; Back cover of paperback edition: (top) Giles Prett/Storey Publishing, (bottom) © Richard Klimesh; Back cover of hardcover edition: Giles Prett/Storey Publishing; Back flap of hardcover edition: © Richard Klimesh
Interior photo credits: © Benson Photography: 95; © Peter Dean/Grant Heilman Photography: 10; © Linda Dufurrena/Grant Heilman Photography: 94 (bottom); © K. J. FitzGerald: 94 (top), 97; © Gemma Giannini: 62, 96; © Gemma Giannini/Grant Heilman Photography: 44, 66; © George H. Harrison/Grant Heilman Photography: 5; © Shelley Heatley: 28, 57; © Richard Klimesh: vii, 37, 40, 41, 56, 61 (top and middle), 72 (bottom), 73, 74 (bottom), 84, 88, 89; © Bob Langrish: 48; © Painet Photo: 93; © Dusty Perin: v, 12, 64, 82; Giles Prett/Storey Publishing: back cover (top), vi (top left, bottom right), ix, 2, 8, 14, 15, 30, 31, 33, 34, 43, 45, 50, 52, 67, 74 (top), 83, 91, 98; © William Shepley: 32, 36, 54, 77 (right), 81, 92, 98; © Arthur C. Smith III/Grant Heilman Photography: vi (bottom left), xii, 38; © Lesley Ward: cover (top right), ii, vi (top right), x, 47, 55, 61 (bottom), 65, 70, 72 (top), 77 (left), 90
Illustration credits: Jim Dyekman: 80, 82 ; JoAnna Rissanen: 3 (left, center left, right), 3 (top), 4, 6–7, 11, 18, 22 (blaze), 23, 24, 27, 32, 35, 42, 49, 51 (1 and 2), 52, 53 (top), 59, 68, 69, 76; Elayne Sears: i (center, center right), 3 (bottom), 12, 13, 14, 16, 17, 23 (all except noted above), 46, 51 (3–5), 53 (braiding sequence), 75, 86, 87

Indexed by Susan Olason, Indexes and Knowledge Maps

Copyright © 2002 by Cherry Hill

Printed in China by C & C Offset Printing Co., Ltd.
10 9 8 7 6 5 4 3 2 1

Library of Congress Cataloging-in-Publication Data

Hill, Cherry, 1947-
   [Horsekeeping]
   Cherry Hill's horse care for kids / by Cherry Hill.
      p. cm.1
Includes index.
Summary: Explains how to choose, understand, handle, feed, groom, shelter and pasture, and care for the health of a horse.
     ISBN 1-58017-407-8 (paperback); 1-58017-476-0 (hardcover) (alk. paper)
   1. Horses—Juvenile literature.   2. Horsemanship—Juvenile literature.
  [1. Horses. 2. Horsemanship.] I. Title: Horsekeeping. II. Title.
    SF302 .H53 2002
    636.1'083—dc21
                                       2001049814

dedication

**To my horse buddies all over the world —
human and equine! Have fun. Be safe.**

contents

a special note to my readers

I am so excited for you! I think that horses are the most wonderful interest you could possibly have. Horses have given me so much enjoyment that I have spent my whole life learning about them. Now I'd like to help teach you how to care for your pony or your horse. Books are great sources of information about horses, but in order to become a really accomplished equestrian, you also need two special mentors.

What is a mentor? A mentor is a wise and dedicated teacher. One of your mentors should be a person who has had a lot of positive experiences with horses and loves and respects them, someone who works safely with horses and can describe to you how and why you should do things a certain way. This person could be old or fairly young. He or she could be a professional horse trainer or riding instructor, your aunt or grandfather, your next-door neighbor, the man who works at the feed store, or your 4-H or Pony Club leader.

If you find a mentor who is willing to share knowledge with you and teach you about horses, you are *very lucky*. Do everything you can do to show your mentor that you are serious and want to learn. Pay attention when he or she is explaining something. Never arrive late or miss a scheduled meeting or lesson with your mentor. Be respectful and polite. A mentor is a great treasure for you to find. Think of things you can do to show your appreciation. Offer to do something that will be helpful, like cleaning a few stalls or soaping and oiling some bridles.

The second mentor you need is a well-trained, trustworthy horse. In Chapter 2, I'll give you some specific advice on how to choose the right horse — one that will be safe and fun for you to handle and ride. Look for a patient, well-trained, experienced horse that has already taught other young people how to ride. Usually such a horse will be old and wise but might not be a beauty contest winner. I've found that no matter what a horse looks like, if he is kind and safe and willing to teach you how to ride, you will love him dearly.

Even though riding is probably the main reason you have your horse, don't forget that your horse depends on you for proper feeding, health care, and exercise. I like to think of riding as my reward for a job well done. There is nothing quite like tacking up a healthy, happy, squeaky-clean horse and going for a ride on a trail or in a show. But to "get there" you have to do your horse-keeping homework first. And that's why I wrote this book especially for you — to help you learn how to take the very best care of your horse buddy.

a note to parents

There is something about a horse that is good for the heart of a child. I can remember that as a preschooler when I had the opportunity to groom and ride a horse, I did not want to wash my hands for fear of losing that wonderful smell. I encourage you to support your child's interest in horses.

The rest of my comments to you are going to sound like a bunch of dos and don'ts. They are! I hope you take my advice to heart so your child's experience will be safe and will add to the development of his or her character.

Most children who are six to seven years old have the motor skills, confidence, and attention to learn to ride a horse safely. If you start your child too young, you could risk frightening him or her. Don't push your child into the show ring. Emphasize wholesome, safe fun before competition and show your child that conscientious horse care and good sportsmanship are the signs of a real winner.

Avoid the green horse/green rider syndrome. Parents often think it would be nice to acquire a foal and let the child and young horse grow up together. However, this arrangement results in a greater chance for mistakes and mishaps. A child needs an older, well-trained, patient, and tolerant horse.

Most suitable children's horses and ponies are between the ages eight and

twenty or even older. Geldings are usually preferred because of their stable dispositions. It doesn't matter if a horse or pony is slightly arthritic as long as he is sound and has an exemplary temperament, very solid training, and good manners.

Horses are not people and should not be treated as people. They need to be treated as horses for their own well-being as well as for your child's safety. Your child should be taught a sensitive but realistic approach to horses. The best way for a child to develop the proper attitude about horses is to focus first on responsible care for horses.

Your child will need to learn many step-by-step routines for handling a pony or horse. If you are planning to teach your own child how to care for horses and ride, think it over again. Even if you are an experienced horse person yourself, it is often better for the child to learn horse care, handling, and riding in a structured program.

Look for a high-quality instructor to coach your child regularly. The 4-H or Pony Club in your area may provide such instruction. You can participate and support your child by organizing meetings, fund-raisers, clinics, and shows and let your child benefit from the group's instructors.

If there are no groups in your area, you might find a professional horse trainer or instructor who can guide your child. To begin your search, ask your county Extension agent for several suggestions and a recommendation. If you know other children who are taking lessons, ask their parents how satisfied they are with their child's instruction program. Call several of the instructors that have been recommended to you and ask how long the instructor has been teaching, if he or she is certified, what style of riding is taught, the cost, and the length of the lesson. If everything sounds acceptable, ask the instructor to provide you with two references. After you've checked the references, you should visit the instructor's facilities and view some lessons in progress. Do this before you entrust your child's safety to a particular instructor.

The more time you spend selecting the best instructor for your child, the better your child's experience will be, and the fewer problems you will have. Become familiar with the various activities available for your child and with the sources of information listed in Chapter 8.

It is essential that you become very familiar with horse behavior. I've seen several instances where a parent contributed to an accident between a child and a horse because the parent panicked, did not know what to do, and therefore either froze and did nothing or did the wrong thing. Also, stay current on your first-aid knowledge and skills. They will come in handy for your child and friends as well as for their horses.

Through all phases of your child's equestrian development, he or she needs safe clothing and tack. You will need to invest in proper footwear and headgear as well as other riding accessories. In addition, your child's horse or pony will require safe, suitable equipment for riding.

You might have heard the phrase "backyard horseman," but that doesn't mean you can keep a horse in your backyard. A small pony will need a minimum of an acre and carefully planned, safe facilities to live in.

Realize that it will probably cost at least $2,000 per year to keep a horse. This figure represents routine costs such as feed, bedding, routine vet care, and farrier care.

Being involved with horses requires an investment of time, money, and hard work. Be sure your child knows that he or she must make trade-offs in order to see that a horse or pony has good care. Sometimes it will be necessary to miss a favorite TV program or a party to take care of a horse's needs.

Depending on the age of your child, you may have to do a portion of the horsekeeping work yourself. I look on this as a bonus for you rather than a burden because caring for horses has given me so much enjoyment. You might find, as other parents have, that after you have cared for your child's horse for a while, you will want to acquire one of your own.

You should know at least as much as your child does about horse care and handling. Refer to the recommended reading list on page 112 for other books that will provide more detailed information on various subjects: training, riding, health care, hoof care, facilities, management, safety, showing, and much more.

I am so grateful to my parents for encouraging me to pursue my interest in horses. Although I didn't own my first horse until I was nineteen, throughout my early childhood my parents created many opportunities for me to learn about horses and ride them. There are many ways you can help your child have a positive and very rewarding horse experience. I hope this book helps you and your young horse-keeper on the way.

chapter one

Understanding Horses

Horses are some of the most interesting and beautiful creatures on earth. If you learn as much about them as you can, the more you will think like a horse and the better horsekeeper you will be.

Horses behave the way they do partly because of how they evolved. For thousands of years before they were domesticated, horses roamed grazing and browsing. They were prey animals who banded together in herds for protection from predators.

Horse Talk

aids
Signals from a rider or handler that tell a horse what to do.

herd-bound
When a horse is too dependent on being near other horses.

pecking order
The order of dominance among horses in a herd.

withers
The part of the horse's spine where the neck joins the back.

Horses are horses. They are not people. Although you may want to kiss your horse to show him how much you like him, he might think you are acting odd and pull away suddenly. Or worse yet, he might nibble your lips to see if there is anything in them to eat, and you could be hurt. Your horse will be most comfortable when you do things that he understands. While he is learning more about you, you will learn more about what it's like to be a horse. For example, you will find that your horse will appreciate a good scratch on the *withers* or neck more than a pat on the nose.

You and your horse can be good buddies, but don't be careless. You must always pay attention, because, just when you least expect it, your horse might suddenly jump sideways and smash your toes. Never fool around when you are handling a horse, no matter how much you trust him.

Horses like to be near other horses. Horses that live in herds may become *herd-bound*. When you try to remove one horse from the herd, all of the horses may become nervous and try to stay near the horse you are taking away. Or, the horse you are trying to take away might stop and refuse to leave the herd. To prevent this, you should not let horses get too attached to each other. A horse that is kept in his own pen or paddock and is handled regularly by you will think of you as his "herd-mate" and will look forward to your visits.

If you have three horses together on pasture, one of them will be the "top" horse in the *pecking order*. When you feed them, the top horse will be the first one to get the hay or grain. That's why it is important to spread feed far apart so all three will get some. Otherwise, the "bottom" horse might not get anything. Sometimes horses bite and kick each other to prove who is the top horse. Don't come between two horses that are determining their pecking order or you might get hurt.

Horses are wanderers.
By nature, horses like to roam around and take a bite of grass here and a bite there. This gives them exercise as well as a way to eat while they are on the move. If you put your horse in a pen or a stall, you must provide him with feed and exercise every day because he can no longer roam and take care of those things himself. If you confine your horse too long, he will become restless. Then, when you finally do take him out, he might be hard to handle because he has so much energy to burn. How would you feel if you had to stay in your room for a whole week!?

Horses would rather run than fight.
If your horse sees something he thinks is dangerous, his instinct is to run away from it rather than face it. Your horse's ancestors survived for millions of years because of this instinct. When a mountain lion approached a horse in ancient times, the horse did not try to fight the sharp teeth and claws; he just broke into a gallop and tried to outrun the mountain lion. That's why some modern horses are so spooky and flighty. They imagine that a piece of blowing plastic or a huge rock might be an enemy in disguise. But even if a horse is afraid of something, his curiosity will usually get

CAUTION If a horse is afraid and cannot escape, he might strike, kick, or crash through barriers, including you, trying to get free.

the better of him, and he will try to figure out what the unfamiliar object is. It might take quite a long time, but eventually a horse will walk up to a suspicious object, sniff it, and touch it with his nose.

Horses have keen senses. A horse keeps a watchful eye on everything around him and immediately notices when something changes. If your grooming bucket has been in one spot for weeks and you move it to the other side of the grooming area, watch your horse's reaction when you lead him into the barn. He might stop for a moment and get a look on his face that seems to say, "What the heck is going on here?" He may even whistle and snort a little while he figures out that the bucket is okay in its new place. A horse's senses protect him and help him identify and locate things.

A horse uses his sense of smell to identify people, other horses, and objects. Mares and foals bond with each other by their individual smells, and horses use smell along with vision to recognize objects and specific people. Your horse will learn to recognize your smell so that even in the dark, he can identify you.

Horses have better vision than you do in many ways. Horses are alert to even small changes around them. If a tiny squirrel moves in a tree quite a distance away, you might not see it, but your horse probably will. Also, your horse can see better at night than you can. But sometimes a horse has a hard time focusing his

eyes to get a clear picture. He might have to raise or lower his head or tilt it to see certain things. Horses have blind spots, places where they can't see things unless they move their heads or bodies.

blind spots

When you are in a horse's blind spot, let him know you're there. Otherwise you could startle him, and he might jump and hurt you. To make sure you don't surprise him, talk to the horse while you touch him: "Hi there, Buck." His blind spots are:

1 the area directly in front of his forehead

2 the area of his back directly behind his head

3 the area directly behind his tail

4 the area directly under his head on the ground and near his front legs

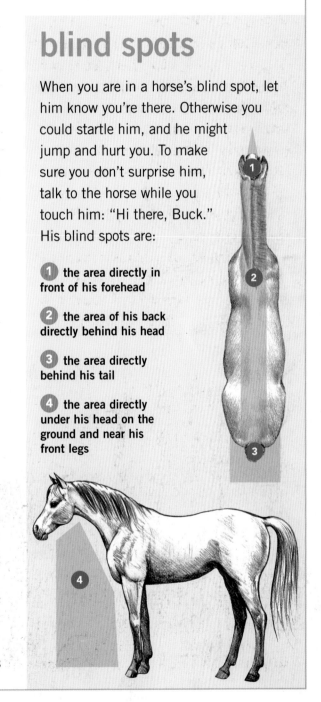

body language

I want to get out of here or **I have a bellyache.** Pawing with front feet.

Hi, pal. Ears tilted forward, head reaching toward you.

I'm getting ready to buck, bite, or kick. Ears pinned back flat against the head.

I wonder what that is over there. Ears forward, head high.

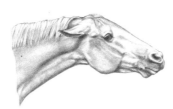

Stay back or **I'll bite you.** Ears pinned back, head reaching toward you.

I'm really concentrating and listening to you. Ears back when you are riding.

Warning: I might kick you. Lifting or stomping one hind leg (not at flies).

I'm irritated or **My stomach hurts.** Swishing tail (not at flies).

I'm afraid or **I don't respect you** or **I'm getting ready to kick you.** Swinging his hindquarters toward you.

Your horse has excellent hearing. His ears can detect sounds above and below the range of sounds you can hear, so he will pick up sounds that you don't even notice. Sometimes a horse will jump when he hears a noisy truck or a high-pitched whistle — sounds that seem perfectly normal to you.

When a horse really wants to know what something is, he looks at it very intently, with his ears pointed toward the object. His ears act like funnels to catch even the faintest sound. Since horses have such good hearing, they can learn to distinguish your voice from other voices. And you don't have to talk loudly or yell at your horse. He can hear you fine when you are speaking in a low or normal voice.

Your horse can easily feel a fly crawling around on the tips of the hairs on his belly. Horses have a keen sense of touch, so you shouldn't use harsh *aids* with your horse. Often just fingertip pressure or a slight weight shift is all that you need. Your horse's nose, lips, and other areas on his head are very sensitive, so be very careful when handling them. Also, whenever you use a bridle on a horse, use light pressure with the reins.

Horses can be "A" students. They have a great ability to learn what we want them to do when we use proper aids. It takes a good horse trainer to make a well-trained horse, but once a horse knows, for example, that when you squeeze him with both legs he should trot, he will remember it for life. The horse's memory is almost as good as an elephant's!

Besides remembering good things, a horse also remembers bad habits. Be sure your horse does not develop bad habits; once he does, it will be very hard for him to "forget" them.

Beware of mares in heat. Once a month during the spring, summer, and fall, when a mare is in heat or "in season," she might have a period during which she is silly, grouchy, spooky, or even mean. Some mares are never grouchy, but if you have a mare that is difficult to handle when she is in heat, ask your veterinarian to examine her. There is a small chance the vet might find something wrong with the mare's reproductive system. Even if there is nothing wrong, it might be best to give such a mare a few days' "vacation" every month when she is in heat.

This pony is saying "What is that over there?"

poll

forehead

eye

crest

bridge
of nose

heart girth

withers

muzzle

upper lip

chin
groove

lower lip

throat
latch

jugular furrow

shoulder

point of shoulder

chest

arm

elbow

forearm

chestnut

knee

cannon

fetlock

pastern

heel

hoof

parts of a horse

Learn the proper terms for the parts
of a horse so you can talk accurately
with your veterinarian and discuss
tack and riding issues with your
instructor and friends.

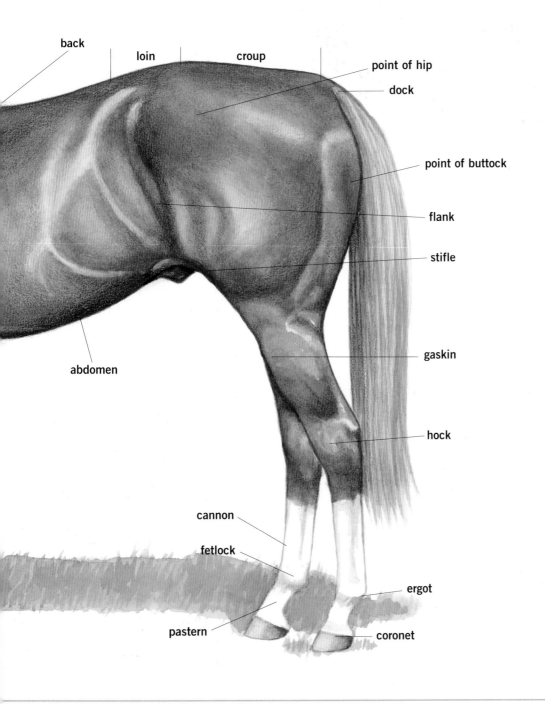

back

loin

croup

point of hip

dock

point of buttock

flank

stifle

gaskin

hock

abdomen

cannon

fetlock

ergot

pastern

coronet

chapter two

Finding the Right Horse

The best horse is a safe and kind horse. When choosing your horse, look for one with good training and manners and a soft eye. Other things to think about are the horse's conformation, size, health, soundness, sex, and age. There are a number of important qualities and characteristics to consider when choosing your horse.

Horse Shopping

Ask your parents to find a well-respected professional horse person in your area who will go horse shopping with you. This could be your regular riding instructor, a trainer, a breeder, or someone else whom you feel comfortable with, who has the time to help you. Together you will look at horses and narrow down the choices to a few horses that you will test ride. Plan to pay for this service.

When you have found a horse that seems to be what you are looking for and is in your price range, schedule a time with the owner for a test ride. Perhaps your advisor will take a test ride first to see if the horse would be suitable for you.

After the test ride, if the horse seems promising, you'll want to have your veterinarian check to see if he has any unsoundness or health problems that would rule him out.

If the horse passes the vet check and any other medical tests required by your state, there will be some paperwork necessary to transfer ownership. This could include registration papers, brand certificate, and other ID verification.

When a very beautiful, spirited horse catches your eye, it may be difficult to look at other less flashy prospects because the beautiful horse has captured your heart. But if such a horse is young, untrained, or poorly trained, and you are inexperienced, it would be better to follow the advice of an experienced horse person. Although you may end up with an older and plainer (but hopefully wiser) horse, the pleasant and safe riding you will enjoy is the important thing. Later, when you are more experienced, you may be ready to progress to a less trained, more spirited horse.

Buying a horse is a big task. Take your time. If you are in a hurry to buy a horse, you might make a bad decision. Plan to take from one month to one year to find your ideal horse. In the meantime, you can learn more about horses by taking lessons.

What to Look For

⊃ Temperament and manners. This should be first on your list of things to check. Look for a horse that is cooperative and calm, alert but sensible. You wouldn't want your horse to jump sideways at every piece of plastic or dog that you pass. Be sure to avoid any horse with the bad habits of biting, kicking, rearing, bucking, or running away. Look for a horse that seems to enjoy people, not a horse that tries to avoid them.

⊃ Level of training. In general, the more training a horse has, the more expensive he will be. If he has a show ring record, that will add to his price. You probably don't need a fancy show horse, but you do need a horse with good overall training so you can safely ride him at all gaits both in and out of an arena. You also need to be able to handle him safely from the ground. Such a horse will usually not be cheap because it takes a lot of time and effort to "make" a good horse.

⊃ Health. A healthy horse does not have a disease or medical condition that requires treatment or keeps him from working. Unless your primary mission is to rescue ill or abused horses, it is not a good idea to buy a sick horse, hoping to nurse him back to health. It is just too time consuming, expensive, and risky. If you find a horse that you really like but he is sick, tell the seller to call you when the horse recovers and you will consider having your vet look at him again.

Your veterinarian should examine the horse before you buy him and can tell you if a horse has a clean bill of health. This is called the "prepurchase exam." Read Chapter 7 very carefully so you become familiar with common horse health problems like founder, navicular disease, heaves, and strangles.

⊃ Soundness. A sound horse does not have a lameness or permanent condition that prevents him from working. You will need an experienced horseperson and veterinarian to help you determine if a horse is sound.

the gaits

A gait is a specific footfall pattern. Most horses are born knowing how to perform the walk, trot, canter, and gallop. Learning the footfall patterns of these natural gaits will help you when training and riding.

Walk. The slowest gait, four beats.

Trot. A two-beat gait between a walk and a lope.

Canter/Lope (left lead). The English term for a three-beat gait; the Western term is "lope".

Gallop (left lead). A very fast canter or lope in four beats; the horse is running.

You need a horse that you can count on every day to perform without pain. Don't buy an **unsound** horse hoping he will get better. Many unsoundnesses are permanent.

Some perfectly sound horses have **blemishes** or scars that might look bad but don't affect how the horse moves or works. Sometimes you can get a good bargain on a horse with a blemish.

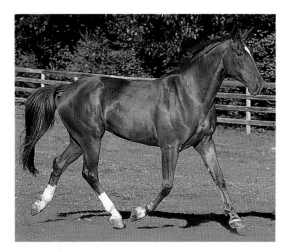

A horse's conformation and type will determine the style and quality of his movement.

⬁ **Conformation and movement.** Conformation is the shape and form of your horse — how he is put together. Generally a horse with good conformation will move well and have less of a chance of developing a lameness. If you want to show your horse, you may have to look for good conformation *and* good looks!

It doesn't matter so much how pretty a horse looks when he is standing still. What is most important is that he moves well and is a comfortable riding horse.

leg conformation

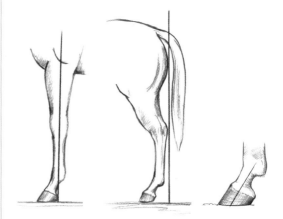

Correct leg and hoof conformation

leg movement

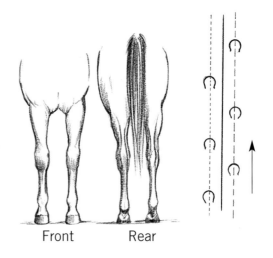

Front   Rear

When horse's leg conformation is correct, travel is straight.

Leg conformation refers to the structure and correctness of a horse's limbs. Good leg conformation is usually associated with a sound horse. Poor leg conformation often leads to lameness.

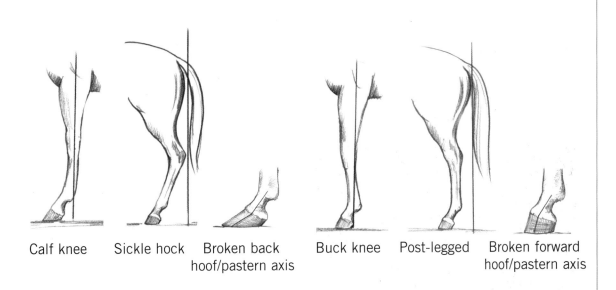

Calf knee   Sickle hock   Broken back hoof/pastern axis   Buck knee   Post-legged   Broken forward hoof/pastern axis

Leg movement refers to how the horse's limbs lift, swing, and land. Good leg movement is straight and efficient. Undesirable leg movement includes stumbling, hitting the opposite leg, and unsightly paddling.

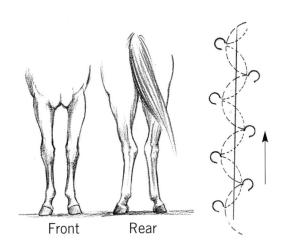

Front   Rear

When horse's leg conformation is toed-out, travel is winging in.

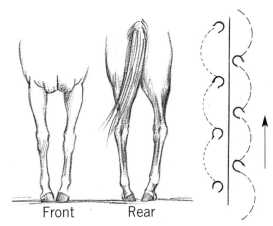

Front   Rear

When horse's leg conformation is toed-in, travel is paddling.

This depends on his body type and conformation as well as his gaits. You'll need help from your riding instructor and veterinarian to help you learn how to evaluate a horse's conformation and movement.

⊙ **Size.** When you are mounted on a horse or pony, your heels should not be below the horse's belly. If your legs are too long for a small pony, you would have no way to give him leg signals — you'd be clicking your heels under his belly!

The horse on the left is too short for its rider's legs, and the one on the right is too tall for its rider's legs.

measuring by hand

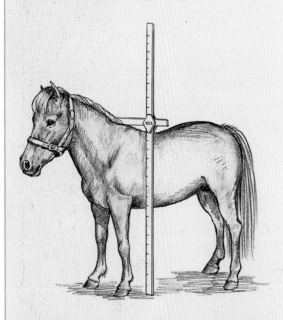

Stand your pony or horse on a level surface. Measure the distance (in inches) from the ground to the highest point of his withers. Divide the number by 4 (4 inches in a hand) to get the height in hands and inches.

An adult pony or small horse will not grow any more, but *you* probably will get taller and grow longer legs. You don't want to outgrow your horse just when you've gotten used to each other. So, with your riding instructor's help, choose a horse that is big enough to last you for a while.

⊙ **Sex.** If you are buying your first horse, I'd recommend a well-trained, quiet gelding, because chances are he will behave the same from day to day. Because of heat cycles, some mares can be sweethearts one day and real grouches the next. So unless you are experienced or know that a particular mare is pretty even-tempered, I'd avoid a mare for a first horse. Under no condition should you consider handling a stallion until you are over eighteen years of age, have the permission of your parents, and have professional guidance. Stallions can be dangerous for you to handle because they are usually full of energy and excitable, especially around other horses.

➲ **Age.** For safety, your first horse should be at least eight years old, but he could be twenty or more. If a horse has been well trained and well cared for, the older he is, the more steady and trustworthy he is. Some horses continue performing well into their twenties. Horses under five years of age are often unpredictable, and their training is not usually solid enough to be trustworthy for a youth rider.

➲ **Pedigree.** A horse's family ancestry might be important if you are purchasing a registered horse and you are looking for certain family lines.

➲ **Color.** A horse's color and markings are often the first thing you see but are the least important when considering safety and reliability. Beauty is as beauty does.

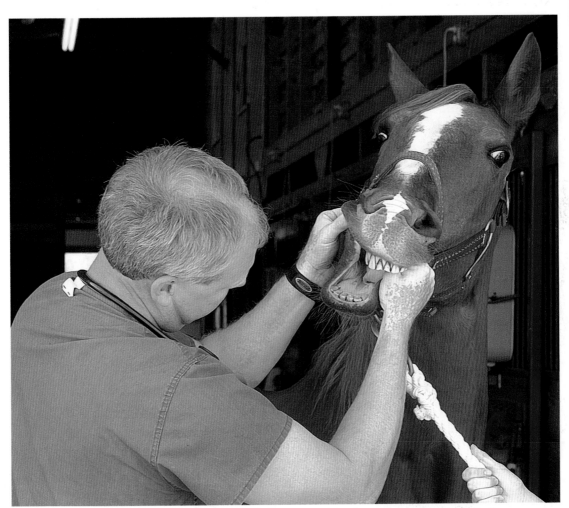

If a horse has no records, your veterinarian can tell the age of the horse by looking at his teeth.

Types and Breeds

Eohippus, the ancestor of today's horses and ponies, lived more than sixty million years ago. Between then and now, horses, classified as *Equus caballus,* have gone through many changes. Along the way, various types of horses evolved, which is why we have different breeds and types of horses and ponies today.

Horses are classified as either heavy horses or light horses. Heavy horses are large horses used for farming and draft purposes. Light horses are those used for riding. Ponies are a subclassification of light horses.

A breed is a group of horses with common ancestors. A breed registry is an association that keeps track of all of the records of the members of a breed. Usually horses in one breed will resemble one another. That's what allows you to look at a particular horse and say, "That's an Arabian" or at a particular pony and say, "That's a Shetland."

Ponies vs. horses. Pony breeds vary in size, color, and gait. Some ponies are more suitable for driving, while other pony breeds make excellent mounts. Horses are generally over 14.2 **hands** in height and are usually described by type or breed. Ponies are generally under 14.2 hands when mature. Mature horses that are shorter than 8.2 hands are classified as miniature horses, not ponies. See page 14 for how to measure your pony or horse.

Horse "type" refers to the kind of work a horse is built to do. If you saw a horse that was well muscled and kept his eye calmly on the cows that were in front of him, no matter what his breed was, you would probably be looking at a stock-type horse. If you saw a tall, lean horse flying like the wind through a pasture, you might be looking at a race-type horse.

Race. Lean but tall horse with a deep heart girth and exceptional speed. Example: American Thoroughbred.

Sport. Large, strong horse used for jumping, cross-country, and dressage. Example: Trakehner.

Show. Horse with flashy, high-stepping gait often used only in the show ring; also called animated. Example: Morgan or Saddlebred.

horse type and use

Although certain horses are versatile and can be used for many activities, some types of horses are better suited for particular uses.

★ **Western Riding:** stock, pleasure, endurance type

★ **English Riding:** hunter, sport, show, pleasure, race type

★ **Trail Riding:** endurance, stock, pleasure type

★ **Jumping:** hunter, sport, race type

Endurance. Small, lean, tough horse that can cover many miles. Example: Arabian.

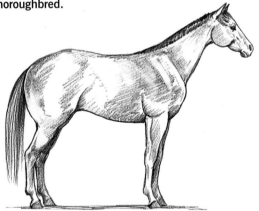

Hunter. Smooth, graceful horse that moves with a long, low reach of his legs. Example: American Thoroughbred.

Stock. Well muscled, quick on his feet, knows cows, has "cow sense." Example: American Quarter Horse.

Pleasure. Balanced, smooth-gaited horse that is easy to ride. Example: individual horses in any breed.

horse and pony breeds

American Saddlebred

★ **Origin:** An American breed established in the 1800s in the Southern states for riding and driving.

★ **Qualities:** Today most Saddlebreds are of the show type with flashy gaits.

★ **Height:** 16 hands

Appaloosa

★ **Origin:** Spotted horse that originated in the northwest United States (Idaho, Oregon, Washington) in the land of the Nez Perce Indians.

★ **Qualities:** An Appaloosa that shows color might be leopard (spotted all over), have a white blanket over his hips with spots in the blanket, or be frosted all over.

★ **Height:** 14.2 to 15.2 hands

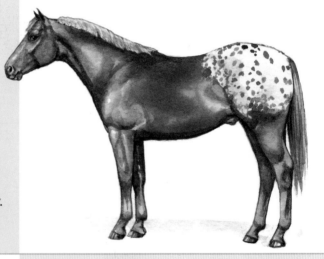

Arabian

★ **Origin:** A breed that traces back to the year 786 in Arabia and has had an enormous effect on other breeds.

★ **Qualities:** The Arabian usually has a beautiful head, a short back, floating action, and often a high-tail carriage. A common color is gray.

★ **Height:** 14.2 to 15 hands

Grade Horse

★ **Qualities:** A grade horse is one that is not registered with a breed association. He might be a purebred without papers, or he might be a crossbreed.

★ **Height:** Variable

Morgan

★ **Origin:** An American breed that traces back to 1795, to a single stallion named Justin Morgan.

★ **Qualities:** Morgans are good horses for driving and riding. They are compact horses with flashy gaits.

★ **Height:** 14.1 to 15.2 hands

Paint

★ **Origin:** Paint horses usually have Quarter Horse and Thoroughbred ancestors.

★ **Qualities:** This stock-type horse has painted body color (large blocks of white and black or white and brown).

★ **Height:** 14.3 to 15.3 hands

horse and pony breeds

Pinto

★ **Origin:** Pinto is a color registry, not a breed.

★ **Qualities:** Coat pattern like a Paint but can be miniatures, ponies or horses of stock, hunter, pleasure, or saddle type. Pinto horses can be any breed such as Quarter Horse, Thoroughbred, Arabian, Morgan, Saddlebred, and Tennessee Walking Horse.

★ **Height:** From miniatures under 8.2 hands to horses over 14 hands

Quarter Horse

★ **Origin:** The oldest and most popular American breed, originating in the 1600s in Virginia. Now the basis for other breeds such as the Paint, Appaloosa, and Palomino.

★ **Qualities:** Well muscled; cow sense; calm disposition.

★ **Height:** 15 to 15.3 hands

Thoroughbred

★ **Origin:** An English breed from the early 1700s that traces to three sires: the Byerly Turk, the Darley Arabian, and the Godolphin Arabian.

★ **Qualities:** The Thoroughbred is a long, lean horse that can cover the ground.

★ **Height:** 15.2 to 16 hands

Warmblood

★ **Origin:** A general term for European breeds of sport horses. Warmblood comes from crossing a hot-blood horse like an Arabian or Thoroughbred with a cold-blood horse like a draft horse.

★ **Qualities:** These animals range in size between a Thoroughbred and a draft horse. Examples are Dutch Warmblood, Selle Français, Trakehner, Hanoverian.

★ **Height:** 16 to 17 hands

Connemara Pony

★ **Origin:** An Irish pony with influence from the Arabian breed.

★ **Qualities:** Connemaras are often gray or black but can be brown, bay, or dun. They are noted for their smooth gaits and make excellent riding ponies.

★ **Height:** Generally one of the taller ponies, ranging up to 14.2 hands

Pony of the Americas (POA)

★ **Origin:** An American pony breed founded in 1956. The breed was developed by cross-breeding Shetland ponies with Appaloosa horses and a few quarter horses and Arabians along the way.

★ **Qualities:** POAs show a wide variety of coat patterns similar to those of Appaloosa horses.

★ **Height:** 11.2 to 13.2 hands

horse and pony breeds

Quarter Pony

★ **Origin:** A breed association for Quarter Horse-type ponies of unknown ancestry; founded in the United States in the 1960s.

★ **Qualities:** Can be any color or coat pattern; must not be gaited; must not show Western gaits only; must have Western stockhorse-type conformation.

★ **Height:** 11.2 to 14.2 hands

Shetland Pony

★ **Origin:** The Shetland Islands, part of the United Kingdom.

★ **Qualities:** The smallest of the ponies, Shetlands come in two types: the Modern fancy driving pony and the Classic sturdy type. They can be almost any color or coat pattern.

★ **Height:** 7 to 10.2 hands

Welsh Pony

★ **Origin:** Wales (part of the United Kingdom)

★ **Qualities:** Welsh ponies are considered to be excellent riding ponies.

★ **Height:** Welsh ponies come in four sizes, ranging from 12 to 15 hands. Three sizes are actual pony sizes (under 14.2 hands), while the larger Welsh size is referred to as a cob (a small horse)

markings

A horse's face and leg markings are his unique "ID Card." It is a good idea to photograph and accurately draw your horse's markings for an identification record.

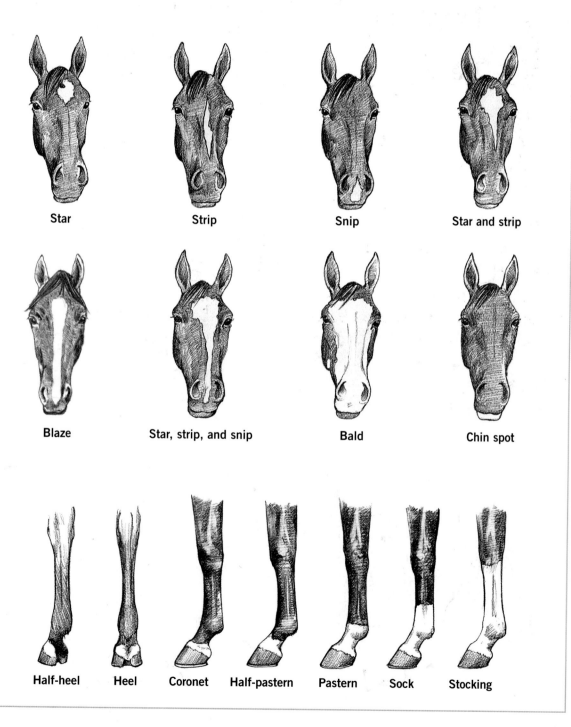

Star Strip Snip Star and strip

Blaze Star, strip, and snip Bald Chin spot

Half-heel Heel Coronet Half-pastern Pastern Sock Stocking

horse colors

Bay

Body color ranges from tan to reddish brown coat with black mane and tail, and usually black on lower legs.

Black

True black over the entire body, including the flank and muzzle, except there may be white leg and face markings. The mane and tail are black.

Blue Roan

A uniform mixture of black and white hairs all over the horse's body. The horse is born this way and stays this color for life. (The color does not get lighter as the horse ages the way the coat of a gray horse does.) The head and legs are usually darker than the body.

Brown

Mixed black and brown hair with black mane, tail, and legs. Often the horse appears black but has light areas around the eyes, muzzle, flank, and inside upper legs.

Buckskin

Tan, yellow, or gold with black mane and tail and black lower legs. Buckskins do not have dorsal stripes the way duns do.

Chestnut

Body, mane, and tail are various shades of golden brown, from sunny gold to reddish brown; some have manes and tails that are the same color.

Liver Chestnut

A very dark red chestnut color, almost mahogany; mane, tail, and legs either same color as body or flaxen.

Dun

Yellow or gold body and leg color, often with black or brown mane and tail; usually has dorsal stripe (stripe down the back), zebra stripes on the legs, and stripes over the withers.

horse colors

Steel Gray

Black skin with a mixture of black and white hairs. The horse is usually born dark (black or charcoal gray) and gets a lighter gray each year until it is almost white.

Dapple Gray

A "middle-aged" gray horse is often dapple gray (small round white splotches) or fleabitten gray (speckles) as his coat gradually lightens.

Light Gray

An older gray horse looks almost white except that his muzzle, inside his ears, and between his hind legs are black. A true white horse (albino) would be pink there.

Grullo

Smoky or mouse-colored body (each hair is this color; the coat is not a mixture of dark and light hairs); mane, tail, and lower legs are usually black; usually has dorsal stripe.

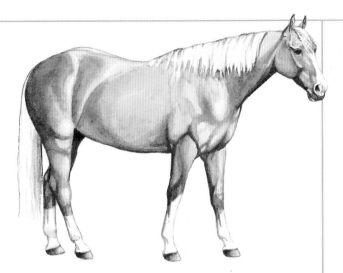

Palomino

Golden coat with a white mane and tail. The coat is sometimes described as being the color of a newly minted gold coin.

Red Dun

Yellowish, light red, or flesh-colored body; mane and tail are reddish, flaxen, white, or mixed; has red dorsal stripe and usually red stripes on legs and withers.

Sorrel

A Western term used to describe a reddish or copper-red body with mane and tail either the same color as body or flaxen.

Strawberry Roan or Red Roan

A mixture of red and white hairs all over the horse's body but usually darker on the head and legs; can have red, black, or flaxen mane or tail.

chapter three

Horse Safety

To be safe when you are handling or riding your horse, start with a safe attitude, dress properly, use safe equipment, and use sensible horse-handling methods. Knowing the best way to handle a horse is your best insurance against injury. Learn some simple safety rules and methods, and put them to use whenever you are working with horses. The best way to gain your horse's trust is to be sensible and safe.

Dos and Don'ts

Follow these safety rules to help you avoid the most common causes of accidents.

◯ Don't take shortcuts when performing everyday chores.
Situation: You spray your horse's head with a hose when bathing him instead of taking the time to wet his head with a sponge. The horse gets water in his eyes and ears.
What might happen: The horse could rear up and land on you.
The remedy: Take your time. Remember, with horses, often the slower you go, the faster you get there.

◯ Do make sure you have enough experience to handle a particular horse.
Situation: You go out to the pasture to catch a horse in a group. The horse keeps avoiding you by standing in the middle of the group of other horses. Every time you try to walk up to him, he turns his rump toward you. The other horses are milling around nervously.
What might happen: You could be kicked or stepped on.
The remedy: When you sense that you're not in charge of the situation, stop right away. Go and get more experienced help.

◯ Don't lose your temper around a horse.
Situation: You are riding your horse and are having difficulty stopping him. You pull harder and harder on the reins and start jerking on the bit. You kick the horse in his sides and say, "You stupid horse!"
What might happen: The horse could rear up and fall over backward on top of you.
The remedy: To keep from getting frustrated, angry, or hurt, stay close to more experienced riders who can help you figure out why you and your horse aren't communicating.

Caution: Hand feeding treats can be dangerous.   **For safety, plan sessions with your buddy.**

◑ **Don't use unsafe equipment.**
Situation: You tie a horse up with an old, weak halter.
What might happen: While you are cleaning his hoof, the horse could lean backward, break the halter, and land on top of you.
The remedy: Make sure all your tack is fit for the job — sturdy, safe, and well-maintained.

◑ **Don't work with a horse that has bad habits.**
Situation: You are mounting a horse that doesn't stand still.
What might happen: He could take off when your left foot is in the stirrup but before you have swung into the saddle. You could be dragged.

The remedy: When riding an unfamiliar horse, ask for assistance and be prepared for anything!

◑ **Do follow your teacher's advice.**
Situation: Your instructor tells you never to ride your horse into the barn, but you think it would be fun and do it anyway.
What might happen: You could hit your head on the doorway.
The remedy: Choose your instructor carefully, then trust her experience and advice.

◑ **Don't work with a young or untrained horse unless you are expert yourself.**
Situation: You are riding a three-year-old that has been started by a trainer and ridden for only one month.

What might happen: The young horse could become confused by your rein signals and start bucking. You could be thrown off and break your arm.

The remedy: Ride well-trained horses until you are an excellent rider. Ride under supervision of the trainer when riding a new horse.

⟡ Don't work in an unsafe area.

Situation: You are riding in a pen with a barbed-wire fence.

What might happen: A small, yapping dog could run into the pen and nip at your horse's heels, or another distraction could occur. Your horse could panic and run into the wire fence. You and your horse could be severely cut.

The remedy: Think ahead and observe potential hazards. Try to fix or eliminate hazards or if not possible, postpone your ride till conditions improve.

⟡ Do learn to read a horse's "body language."

Situation: When you try to catch your horse, you reach for his nose.

What might happen: He could turn away, swing his hindquarters toward you, and step on your foot.

The remedy: Learn where a horse's *blind spots* are and where he *likes* to be touched.

⟡ Don't show off or be silly.

Situation: You are showing your friend how quiet and sweet old Buck is by wrapping his lead rope around your waist.

What might happen: Some botflies could suddenly go into Buck's nose or something else could startle him. He could rear and take off, dragging you behind, tangled in the rope.

The remedy: The best things to "show off" are your excellent horse handling skills and horsekeeping habits.

⟡ Do learn safety practices to help you deal with bad luck.

Situation: It is a muddy day, and you are trotting your horse down a small hill that is normally safe to trot or canter down.

What might happen: Your horse could lose his footing and his balance and fall.

The remedy: Use safe stirrups and riding boots and learn how to dismount safely from your horse in an emergency.

When leading, stay in proper position next to your horse's shoulder.

Wear gloves, boots, and a helmet to be safe when working around horses.

Safe Clothing

Proper clothing is an important part of safe handling and riding. Always wear hard-toed boots or shoes when you are working around your horse. If you wear soft shoes or sandals and your horse accidentally steps on your foot, your toes could be seriously hurt. Also, be sure that your boots or shoes have soles and heels that give you good grip on the ground. If you wear slippery shoes around your horse, you can lose your footing and slide underneath him, which puts you in a dangerous position. When riding, always wear a boot with a heel to prevent your foot from slipping through the stirrup.

Because you will need to hold onto a lead rope when handling your horse, you should get used to wearing gloves when you are working with your horse. If your horse suddenly darts to the side and the lead rope "zings" through your hand, it could leave a nasty rope burn behind. The heat from the fric-tion of the rope moving so fast against your skin is as painful as a burn from a hot stove. Gloves will protect you in such a case.

Wear a safety helmet whenever you ride or work around horses. A helmet will protect your head if you fall or if a horse bumps into you or hits you with a hoof. Don't think that a safety helmet is necessary only for jumping. You should wear one for all types of riding and when you are handling horses from the ground, too.

If you are going to show, check the rules of the association governing the show to see what type of helmet is required.

You can buy many types of helmets, including ones that look like Western hats. Not all, however, are constructed to absorb the concussion from a fall. Helmets approved by the Pony Club, the American Horse Shows Association, or the United States Combined Training Association will be labeled ASTM/SEI to show they have passed strict tests that prove they will protect your head.

Be careful when buying a used helmet, because you won't know if the helmet has been damaged in an accident. A used helmet could look perfectly fine even though the inside protective layer may have been damaged and won't protect your head in case of a fall. Ask your instructor to help you choose a safe, approved helmet and then wear it whenever possible when you are handling or riding horses.

Safe Equipment and Facilities

Use tack and equipment of the strongest type, and inspect it regularly for wear. Tack should be well stitched and constructed from durable materials. Choose a well-made rope or nylon wide-web halter for everyday leading and tying. Use a ⅝-inch- or ¾-inch-thick cotton lead rope with a sturdy snap on it. Be sure your tack is not so old and worn out from dirt, sweat, rain, hot sun, or long-time use that it is no longer safe. When choosing tack, be aware that some tack is made to be very colorful and attractive but might not be the strongest and safest tack to use.

Your training facilities should also be strong and safe. The place where you tie

your horse needs to be stout. It is best to tie your horse to the post of a specially designed tie area. If you tie to a rail or board, your horse could pull back, remove or break the board, injure himself and others, and possibly panic, dragging the board along with him. If this happens to your horse only once, he will be suspicious every time he is tied.

Training pens and arenas for horses should be at least six feet tall and strong. When you longe or ride a horse in a training area, you want to be sure he won't try to get away. A frightened or bolting horse might try to go over or through a flimsy or low training arena fence.

Clean and inspect your tack regularly to be sure it is safe and ready for riding.

Safe Handling

Here's where you put your safe attitude to use — by understanding the horse's way of thinking and by planning ahead to avoid accidents.

Approaching a horse. Always speak to a horse as you are approaching him. Approach at an angle, aiming for the horse's shoulder. Never approach a horse by aiming directly at his head or his tail. Remember his blind spots.

Touch a horse first by placing a hand on his shoulder or neck. Don't pat the end of his nose. Doing so might cause him to move away, or it might encourage him to nibble.

Either walk around a horse well out of kicking range or move around the horse by staying close, with your hand on his hindquarters to let him know you are there. Never walk under or step over the tie rope.

haltering basics

Use proper haltering procedures to develop good habits in your horse and avoid accidents. Don't leave a halter on your turned-out horse, as he may hook it on a post when rubbing or on his own hind shoe when scratching his head with his hoof.

1 Approach the horse from the near (left) side and hold the unbuckled halter and rope in your left hand. Reach under the horse's head with your left hand. Reach over the neck and take the rope with your right hand.

2 Make a loop around the horse's throat-latch and hold the loop with your right hand. If the horse tries to move away at this stage, you can pull his head toward you with the loop of the rope.

3 Next, hand the halter strap with the holes in it under the horse's neck to your right hand which is holding the lead rope loop. With your left hand, position the noseband of the halter on the horse's face.

4 Then bring your hands together to buckle the halter. Remove the rope loop from around the horse's throatlatch. You're ready to lead.

Leading a horse. Use an 8- to 10-foot lead rope to lead a horse. When leading from the left, with your right hand, hold the lead rope about 6 inches from the snap that is attached to the halter. With your left hand, hold the balance of the rope in a safe configuration such as a figure eight.

Don't hold the lead rope in a coil in your left hand. If your horse suddenly pulls, the rope might tighten around your left hand, which could become trapped in the tightened coil.

When leading, if necessary use your right elbow against the horse's neck to keep him straight and to prevent him from crowding you. Make the horse walk beside you. You should be next to his neck or

Although this pony is leading on a fairly slack line, the handler is getting a little ahead of the pony.

safe turn out

Teach your horse patience when turning him loose. Do not let him bolt away. Before you remove your horse's halter, drop a horse treat on the ground. Apply the loop around the horse's neck, remove the halter, and then release the loop. You should hold the horse momentarily with the loop and then gently push him away from you with your right elbow. Your horse will put his nose down to the ground for the treat. The neck stretch and chewing both relax a horse and help him forget about bolting away.

shoulder. He should not lag behind or pull ahead of you. Turn the horse away from you and walk around him rather than having him walk around you. Turning away is the safest way to turn. Once you know your horse, you will want to be able to turn him toward you, too.

Work your horse from both the right and left sides so that he develops suppleness and obedience each way and does not become one-sided. If a horse resists or balks when you try to lead him, do not get in front and try to pull. Instead, stay in the proper position at the shoulder and urge the horse forward with a light tap from a long whip held in your left hand.

Never wrap a rope or strap around your hand, arm, or other part of your body. If a horse spooks suddenly and bolts, you may be unable to free yourself from the horse and could be hurt badly.

Tying a horse. When you need to tie your horse, be sure he is wearing a strong nylon rope or web halter and that the lead rope and hardware are heavy duty.

✅ Know how to tie the quick-release manger knot without hesitation (see steps below).

✅ Keep your fingers out of loops when tying knots.

✅ Be sure your horse is well accustomed to being tied in other ways before you attempt to tie him in a *cross-tie.*

✅ Tie horses a safe distance away from each other.

✅ Always tie to a strong post or tie ring, not to a rail that may be pulled loose.

✅ Never tie a horse with bridle reins. It would be far too easy for the reins to snap, which may cause damage to the horse's mouth and increase the potential for the horse to develop the bad habit of pulling when tied.

✅ Always tie at the level of the withers or higher. Tie to a solid post or tie ring, never to a flimsy or loose rail, a portable panel, or a wire fence.

✅ Untie a horse and hold him temporarily with the lead before removing the halter. Never remove the halter while the rope is still tied to the post. These two practices will help prevent your horse from developing the bad habit of pulling away.

tying the quick-release knot

Run the rope through the ring. Hold the rope in your left hand.

Make a fold (bight) in the rope and cross it over your left hand.

Pass the bight through the circle you've made.

Pull the bight through until about 6" long. Snug the knot.

chapter four

Feeding and Nutrition

Eating is very important to your horse. Just show up an hour late for chores and you'll hear about it! Horses evolved as wandering grazers, digesting small meals of low-nutrient foods all day. You can see why eating one or two large rich meals a day could upset a horse's digestive system. Understanding just what your horse needs, and why, is necessary to develop a proper feeding program for him.

12 Rules of Feeding

1 Know how much your horse weighs. You can't feed your horse properly unless you know how much he weighs. Since your horse is too big for your bathroom scale, you need another way to determine his weight. If you measure his heart girth (see page 42) with a special horse weight tape, it will tell you how much he weighs.

2 Know how many pounds of hay your horse needs. Your horse feels best when he is fed a high amount of bulk (hay) and a low amount of concentrate (grain). Be careful not to feed too much grain.

For every 100 pounds your horse weighs, feed him about 1.5 pounds of hay per day. So, if your horse weighs 1,000 pounds, he should receive 15 pounds of hay per day (1.5 pounds of hay x 10). Since you will probably feed him twice a day, split his ration in half. A 1,000-pound horse would get 7.5 pounds of hay in the morning and 7.5 pounds in the evening.

Weigh the hay at each feeding because flakes of hay can vary. One might be 2 pounds and the next one 7 pounds. If you simply feed your horse "two flakes of hay" you really don't know how much you are feeding him.

3 Know how many pounds of grain your horse needs. Not all horses require grain in their diets. Grain should be fed to young, growing horses, horses in hard work, and lactating broodmares (mares that have nursing foals). Feed each horse his or her grain individually according to the horse's specific needs. This avoids competition and fighting and will keep some horses from gulping and getting too much while others get very little.

Grain should be fed by weight, not volume. Don't use a scoop to measure grain unless you know exactly how much

feeding your horse

Here is a quick and handy summary of the twelve rules of feeding your horse. To provide the very best care for your horse, memorize and follow this list.

- ★ Know how much your horse weighs.
- ★ Know how many pounds of hay he should be fed.
- ★ Know how many pounds of grain he needs.
- ★ Don't overfeed your horse.
- ★ Don't underfeed your horse.
- ★ Know what mineral supplements your horse needs.
- ★ Feed your horse at least twice a day.
- ★ Feed him at the same time every day.
- ★ Be sure he always has fresh water to drink.
- ★ Make any changes to his feed gradually.
- ★ Introduce your horse to pasture grass gradually.
- ★ Never feed or water a horse that is hot from exercise.

a particular scoop of a certain grain weighs. Instead of feeding by the can or scoop method, feed your horse by weighing his grain. First weigh the empty can and write down how much the can

weighs — for example, 1.2 pounds. Then fill the can with grain and weigh it. If the can with grain weighs 4.2 pounds that means there are 3 pounds of grain inside the can.

Compare how much the same size can would hold of bran, sweet feed, oats, pellets, and other horse feeds (see box on page 41). Oats are much lighter than corn, for example, so a can of oats will weigh far less than the same can of corn.

4 Don't overfeed your horse.
Although it is tempting to show your horse how much you love him by giving him extra feed, often this is the worst thing you can do. He might *founder* or suffer from colic. When a horse feels sick, he cannot throw up. He will have to wait until the feed that is making him sick passes through his digestive system.

To prevent your horse from accidentally overeating, be sure your grain room has a secure lock on it. If your horse were able to get into the grain room, he would probably "eat himself sick" and suffer from colic or founder.

Overfeeding can also make your horse gain too much weight. A horse that is overweight has more stress on his legs, which could contribute to his becoming lame.

whole oats   sweet feed   pelleted grain ration

5 **Don't underfeed your horse.** If you don't feed your horse enough, he could grow too thin and weak, be cold all the time, lack energy, and be unable to ward off sickness.

6 **Know what mineral supplements your horse needs.** Balance your horse's feed ration by providing free-choice trace mineralized salt, which contains sodium chloride (salt) and usually iodine, zinc, iron, manganese, copper, and cobalt.

Depending on the horse's age and the type of hay he is getting, he might need additional calcium and phosphorus. The best way to be sure your horse is getting what he needs is to provide him with free-choice access to a calcium and phosphorus mineral block that is made specifically for horses.

However, if you are feeding only grass, hay, and grain to a very young horse (under the age of two), you might need to provide him with an extra source of calcium, because both grass hay and grain have more phosphorus in them than calcium. On the other hand, if you are feeding a young horse only alfalfa hay, which is very rich in calcium, you might need to add some phosphorus to his diet to balance the minerals. If you need help evaluating the hay and grain you are feeding, ask your veterinarian or county extension agent, who can either evaluate it for you or tell you where you can get your hay tested.

7 **Feed your horse at least twice a day.** Since horses evolved as wandering grazers, their digestive systems are geared

how much does it weigh?

Each grain has its own weight. When you fill a quart-sized can with one of the following grains commonly fed to horses, here is how much it will weigh:

bran   = .5 pound
oats   = 1.0 pound
barley = 1.5 pounds
corn   = 1.75 pounds

to many small meals each day. That's why you should feed your horse two to three times each day.

Tip: If your horse gulps his grain as soon as you give it to him, feed him his hay first to take the edge off his appetite.

how to weigh your horse

If you don't have a horse weight tape, you can use an ordinary measuring tape and find out approximately how much your pony or horse weighs by measuring his heart girth (hg) and body length (bl) and using this formula:

$$\frac{hg \times hg \times bl}{330} = \text{body weight}$$

Note: Measure your horse's body length in a straight line from the point of the shoulder to the point of the buttocks (see pages 6–7, parts of a horse).

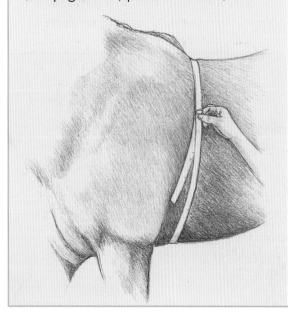

If he is still hurrying, add some rocks about the size of golf balls to his grain to slow him down. Feeding him his grain in a large, shallow pan will force him to eat more slowly than if you put his grain in a small, deep bucket that allows him to gobble big mouthfuls.

8 **Feed your horse at the same time every day.** Horses have an extremely strong internal clock, especially when it comes to feeding. Feeding late or inconsistently can result in colic and other digestive upsets. Be sure to feed the same amounts at the same time every day.

9 **Be sure your horse always has fresh water to drink.** Always make sure your horse has good quality, free-choice water. He may not always drink when it's convenient for you, but he will drink as part of his daily routine, usually an hour or two after eating hay. If your horse does not get enough water, he can lose his appetite or suffer a bout of impaction colic.

In winter a horse should not be expected to eat snow or ice for water. The best routine in winter is to draw fresh water for your horse every day. If you live in a cold climate and your horse is out on a pasture with a creek, you may need to break the ice to keep the water hole open so your horse can drink.

10 **Make any changes to his feed gradually.** Whether it is a change in the

Feed your horse at least twice a day and at the same time every day. Most of his ration should be hay.

Feed grain sparingly, and make changes to your horse's diet gradually.

type of feed or the *amount* being fed, make only small changes. Maintain the new level for several feedings before making another small change. For example, if you are feeding 2 pounds of grain per feeding and want the ration to be increased to 3 pounds per feeding, increase to 2½ pounds per feeding and feed that for at least two days. Then increase to 3 pounds.

If you are making a change in hays, start by replacing one-quarter of the "old" hay in the ration with new hay. Feed this combination for two days. Then increase the amount of new hay so that the hay ration is one-half old hay and one-half new hay for two days. Then

feed one-quarter ration of old hay and three-quarters ration new hay for two days. Finally, feed all new hay.

If you feed your horse 2 pounds of grain or more per feeding and you have not exercised him for a few days, be careful when you start him up. Warm him up very slowly. If you suddenly climb on him and gallop off, his muscles might cramp and cause him severe pain and even damage. This is called "tying up" (see Chapter 7). To prevent tying up, decrease your horse's grain ration if you know he will not be working for a particular period. Then, when he is back to regular work, return his ration gradually to its regular level.

11 Introduce your horse to pasture grass gradually. When turning a horse out to pasture for the first time, first be sure he has had a full feed of hay. Limit his grazing time to one-half hour per day for the first two days. Then he can be on pasture one-half hour twice a day for two days, then one hour twice a day, and so on, until he is out all day.

12 Never feed or water a horse when he is hot from exercise. Do not feed a horse immediately after hard work, and don't work a horse until at least one hour after a full feed. The horse has a sensitive digestive system, and you must be considerate of his work requirements around meal times.

When a horse is hot from exercise, let him take only very small sips of water. If you have just come back from a trail ride and your horse is breathing hard, walk him for a few minutes, then let him take a few sips of cool, not cold, water. (Very cold water could be too much of a shock to his system.) Then walk him for a few more minutes. Offer him a few more sips of water, and so on. When his breathing is normal and he has cooled down, you can feed him some grass hay.

A horse that eats too much rich or green feed can quickly become overweight or suffer gaseous colic or laminitis (founder). Keep a close watch on horses that are on pasture.

Nutrients

Make sure the rations you provide contain the nutrients your horse requires. The five nutrient groups your horse needs are water, energy, protein, minerals, and vitamins.

Water, clean, fresh water, should be available to your horse twenty-four hours a day, every day. Don't expect your horse to quench his thirst by eating snow; it would take him a long time to get enough water that way, and it would also make him cold inside.

Your horse will drink more water than normal if:

- the weather is very hot or humid
- he is exercised hard
- he is eating a lot of salt
- he is eating alfalfa hay
- she is a lactating broodmare

Caution: Feeding two horses together can result in fighting, injury, and one horse going hungry.

Carbohydrates (starches and sugars) in hay and grain give your horse energy. How much energy feed your horse needs depends on his age, his weight, and his level of activity.

Protein is necessary for all horses but especially for young growing horses. A horse may not grow properly if he isn't fed enough of the right kind of protein.

On the other hand, if he is fed too much protein, his bones may grow abnormally. Also, the excess protein will have to be expelled in his urine. A horse that is fed too much protein will drink more water than normal and urinate more than normal. Feeding too much protein is not only wasteful, but it may also be dangerous for a young horse and hard on any horse's kidneys.

group feeding

If you have three horses on pasture, place the feed in four or more piles that are far enough apart from each other that the horses won't fight over them.

tying a hay net

When tying up a hay net, run the tie string through the tie ring **1**, then through the bottom of the hay net **2**, and back up through the tie ring **3**. The hay net should be tied high enough so that when it's empty the net does not hang so low that the horse could get his legs tangled in it.

Minerals are important for many of your horse's body functions. All horses should have free-choice trace mineral salt. "Free choice" means that your horse should be able to get to the mineral salt at all times and eat as much as he wants.

Many horses need extra calcium and phosphorus in their diets as well, so you should consider purchasing a trace mineral salt block that has 12 percent calcium and 12 percent phosphorus added to it.

Vitamins are present in the hay and grain a horse normally eats, so you rarely need to add vitamins to your horse's feed. Horses over the age of twenty-five, however, often benefit from the B vitamins and vitamins C and E.

Types of Feed

Horses are customarily fed hay, grain, and minerals. Hay can be fed as "long stem" hay as in bales or as cubes or pellets. Grain comes in a variety of forms — whole grains, processed grains, sweet feed, and pellets. Minerals can be fed loose or in block form.

Hay. A horse's digestive system is designed to digest bulky feed like hay and pasture. Grass is the traditional horse hay and includes timothy, brome, orchard grass, and other grasses. When it is baled properly, it makes good horse feed, but if it is too mature, it's not much better than straw, nutritionally. Good hay is free of mold, dust, and weeds and has a bright green color and a fresh smell. It is leafy, soft, and dry but not brittle. There is no dampness that could cause spoilage or molding.

Alfalfa hay is higher in protein than grass hay and has three times the amount of calcium and many more vitamins than grass. Alfalfa is often used to feed growing young horses, because calcium is good for bones. However, alfalfa has so much calcium that it can cause problems with bone growth. Check with your veterinarian before you consider feeding alfalfa.

A grass-alfalfa mix hay that contains up to 20 percent alfalfa often makes the best horse hay.

Pasture provides necessary exercise and nutrients for horses. To make best use of a pasture, let your horse eat the pasture grass when it is 4 to 6 inches tall. After he has grazed it down, move him to another pasture.

Native pasture is what appears on the land without extra seeds having been planted by humans. When you drive through parts of Wyoming, for example, all you see is native pasture. Improved pasture is a field that has been planted with specific pasture-grass seeds, and improved pastures are usually watered or irrigated. If you have improved pasture, one or two acres could support two horses during the six-month grazing season of spring, summer, and fall. However, it may require fifty or more acres of native dry range land to support a single horse.

Grain. Many horses don't need grain. Young horses, working horses, pregnant mares, and mares with foals usually require grain.

Oats are the traditional horse grain. They provide fiber (from hulls) and energy (from kernels). Oats are the safest grain to feed a horse, because they are a good balance between concentrate and roughage.

A horse eats 10 to 20 pounds of hay per day.

Corn has a very thin covering that does not have much fiber, but the kernel does provide a great deal of energy. Because corn is so quickly and easily digested, it is too concentrated for some horses — a can of corn has twice the energy content of the same-sized can of oats.

Commercially prepared horse feeds are available as pellets or grain mixes. Pellets might have both hay and grain in them. "Sweet feed" grain mixes are usually made up of oats or barley and corn, molasses, and a protein pellet. Feed companies create different products for different horse groups — you can find special feed for foals, yearlings, performance horses, broodmares, and even senior horses. When you feed commercial grain mixtures, be sure to weigh carefully each horse's amount on a scale; do not carelessly give your horse a scoop of feed without knowing how much the grain weighs.

Mineral and protein blocks provide

nutrients that supplement your horse's main diet of hay and grain.

Salt blocks that are pure white provide sodium chloride, the same salt you use on your table. These blocks are fine for horses who receive a mineral supplement in their feed.

Trace mineral salt is regular table salt with important minerals added and usually comes as a red block. Your horse should be able to get to a trace mineral block at all times and eat as much as he wants.

Put a trace mineral salt block where your horse can get to it any time.

A molasses protein block is usually brown; contains salt, molasses, and protein; and is like a big candy bar for horses. Many horses will gnaw down one of these blocks in a few days if you don't roll it out of his reach. Use a protein block only if the hay is very low in protein. Then you will have to watch that your horse doesn't eat too much of the block at one time. If he gobbles half a block in one day he might develop colic or diarrhea or become dehydrated from eating too much molasses or salt.

Calcium and phosphorus are very important minerals for all horses. The amount a horse needs depends on what kind of hay you feed and how old the horse is. Most of the time a horse's calcium and phosphorus needs can be met with a 12 percent calcium and 12 percent phosphorus block (often called a 12:12 block). Ask your veterinarian and feed store owner for help in deciding what type of block your horse needs.

Feed Storage

Grain should be stored in a mouseproof, locked grain room. If your grain room is not tight enough to be mouse-free, then store your grain in metal or heavy plastic barrels and let your cat roam around the feed room to keep the mice away.

Before storing hay, be sure it is fully cured and dry. Ideally, hay should be stored indoors, stacked on wooden pallets to keep it off the moist ground. If you must store hay outdoors, stack it very tightly on pallets, and cover it with a tarp. And be sure the tarp has no holes. A tarp with holes is worse than no tarp at all.

feeders

It is natural for a horse to eat at ground level. But don't feed a horse on sandy or loose soil. If a horse eats sand or dirt along with his feed, he can suffer sand colic, which results from a large amount of sand blocking the horse's intestine. The sand slows down and clogs up the horse's digestive system (see page 86). Instead, feed him his hay on a clean matted area.

Be sure feeders are clean and safe. Don't let feed accumulate in the bottom of feeders. Moldy or spoiled feed can create problems for your horse and mean large veterinary bills for you. Routinely check all feeders for sharp edges, broken parts, loose wires, nails, or any other hazard. If you use hay nets, tie them securely and high enough so your horse can't get his leg caught in the net.

pasture
hayrack

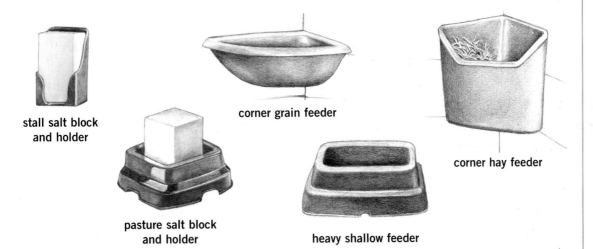

stall salt block
and holder

corner grain feeder

pasture salt block
and holder

heavy shallow feeder

corner hay feeder

chapter five

Grooming and Bathing

Grooming serves many purposes. It cleans your horse's coat by removing dirt, sweat, dead skin cells, and loose hair. It warms your horse up mentally by letting him grow used to your touch. It warms him up physically by increasing his circulation. It brings natural oils from the skin to the surface to make his coat shine. It lets you check his body for nicks or bumps or sore spots. It helps you accustom your horse to overall body handling so he isn't ticklish. And it's good for your relationship with your horse.

Grooming Your Horse

Here are the steps in grooming a horse, with the tools used at each step. Every time you ride your horse, give him a thorough grooming before you saddle up. If you keep him blanketed, the grooming routine will take less time.

1 Use a **hoof pick** to remove mud, manure, stones, and splinters from the frog and sole. One type of hoof pick has a stiff bristle brush on the other side for removing mud from the hooves.

2 Use a soft or "gummy" **rubber curry** to loosen up mud and sweat on your horse's coat and to get shedding hair to fall out. The soft rubber "nubbins" stimulate the skin to release hair oil to make a shiny coat. Use a curry in a vigorous circular motion. The small rectangular type is easier for small hands to hold.

3 Use a **rubber grooming mitt** for the same purposes as the rubber curry but on sensitive areas such as your horse's legs and head. Put it on like a glove and then use your hand to conform to the part of the horse's body that you are working on.

4 Use a stiff-bristled **dandy brush** (sometimes called a "mud" brush) to remove the large pieces of dirt and hair that you brought to the surface of the coat with the rubber curry. The dandy brush is used with a short stroke and a flicking motion of your wrist. Rather than just push the dirt and old hair along the coat from front to back, you want to flick it off.

5 Use a **body brush** next. A body brush is a short, soft-bristled brush that further cleans the coat after the dandy has been used on the body and the rubber mitt has been used on the head and legs.

When grooming ticklish areas, such as your horse's flank, place your other hand firmly on his hindquarters to reassure him.

6 Use a **stable cloth**, a section of an old bath towel, or a wash cloth, either damp or dry, to clean your horse's eyes, nostrils, anus, udder, and sheath.

7 Use your **fingers** to pick through a horse's tail or mane. Sometimes it helps if you oil your hands a little (use one teaspoon of baby oil or a mane and tail product especially made for horses). When brushing is necessary, use a regular human hairbrush, not a comb. Combs cause too much hair to be pulled and broken.

Begin brushing a tail or a long mane from the bottom, and as you get the tangles out of the ends, you can work your way up.

It is not good to brush and comb the mane and tail every time you ride. If you do, your horse will soon have a thin mane and tail! Instead, finger through the mane and tail daily and brush it out completely every week or so.

8 Use a **tightly woven cloth** to remove any remaining dust from the coat and smooth the hair. Always wipe in the direction of hair growth.

A metal **curry comb** is not designed to be used on the horse's body. It is for keeping the bristles of your dandy brush clean. Hold the metal curry in your left hand while you are brushing with your right hand. Every time you make five strokes or so with the dandy brush in your right hand, run the bristles of the brush over the teeth of the metal curry to clean the bristles.

braiding basics

For competition purposes, horse and rider must present a certain appearance, depending on the class. For many classes, the mane and forelock are braided while the tail is left loose, but for many hunter competitions, the tail must also be braided. Never braid the tail without also braiding the mane and forelock.

braiding the mane

1 Make a regular three-strand braid, starting with the right section over the center one, and the left over the middle. At the end of the braid, bring a folded piece of string behind the braid and loop the ends through the folded middle.

2 Poke the ends of the string through the top of the braid and pull snugly, folding the braid in half.

3 Separate the strings, then loop them around the top of the braid and tie off with a square knot.

braiding the tail

1 Starting with damp hair, separate three narrow strands from the sides of the tail and begin to braid.

2 Continue braiding, bringing new strands in from the sides at each step. When you reach the end of the dock, stop adding new strands and just finish off the braid with the remaining hair.

3 Fasten a small rubber band at the end and use a needle and heavy thread to pull the end of the braid under the "wrapped" section, bringing the needle out a few inches above the resulting loop. Tie off and cut the thread.

Bathing

Bathing a horse is a big job and must be done right for the horse's health and for both horse and human safety. It's worth postponing a bath until you can round up an assistant.

A horse should be bathed often enough to keep him clean but not so frequently that his hair and skin become dry. The horse's skin produces a valuable oil that repels water, lubricates the skin, keeps bacteria and fungus at a low level, and makes the coat shine. But the same oil also attracts dirt and causes it to stick to the coat. You need to let the skin oil do its job, but you don't want it to become too heavy in your horse's coat. Too much bathing can remove a horse's natural protection and allow his skin to become dry and populated by harmful skin organisms. Frequent baths can also cause leg and hoof problems. (See information about sanitation on page 82.)

When and where to bathe. It should be safe to give your horse a bath when the temperature is above 50°F if you can bathe and dry him out of the wind.

When bathing a horse for a show or photos, bathe him the day before and keep him blanketed. A freshly shampooed coat "stares," meaning the hairs stick straight out from the horse's body. But after

It's a good idea to have an assistant hold your horse while you give him a bath.

twenty-four hours under a blanket, the skin oil makes the hair lie flat and shine.

If you have a wash rack indoors, that is the ideal place to bathe. The floor of the wash rack should be texturized so the horse won't slip.

If you will be bathing your horse outdoors, choose a place that won't get muddy; otherwise, it will be hard to do a good job. A concrete pad or a rubber-matted area is best. Of course, it needs to be close to a water faucet.

Bath day. On the day you plan to bathe your horse, prepare by getting your supplies and equipment ready.

If you don't have warm water in your barn, fill three or four five-gallon buckets with cold water and set them out in the sun to warm for a few hours. Warm water is more pleasant for your horse, and it does a better job of cleaning his coat because it dissolves oils and dirt more effectively than cold water does.

While the water is warming, gather your bathing equipment: a clean halter, wool cooler, hose, hose brush, buckets, sponges, cloths, rubber mitt, leg brush, sweat scraper, large towels, shampoo solution, and conditioner solution.

What kind of shampoo should you use? Choose a well-known horse shampoo. (Ask your instructor, someone at your farm supply store or tack shop, or another experienced horseperson to help you buy a good shampoo for your horse.) Dish detergent and laundry detergent are too harsh. They

bath time

Two or three baths a year is about right for most horses: a late spring bath just after your horse has finished shedding; a midsummer bath; and a fall bath just before your horse starts growing his winter coat but while the weather is still warm.

will dry out your horse's hair and could irritate his skin. Use a mild shampoo and dilute it according to the following directions. Take a large, clean plastic squeeze bottle (from ketchup or dish soap) and fill it with water up to about an inch from the top. Squeeze two good squirts of your horse's shampoo into this bottle of water. This is what you will use as shampoo for your horse. If you put straight shampoo on your horse, you will never be able to rinse the soap completely out of his coat.

shampooing your horse's head

It is best to wet and rinse a horse's head using sponges full of water instead of a hose. That way you can be more careful about not getting soap and water in his eyes, ears, or nose. If you spray your horse's head with a hose and get water in his eyes and ears, he might learn to hate bathing. To rinse his head properly, hold a sponge full of water at his forelock and slowly squeeze, letting the water trickle down his face. Once his head is rinsed really well, you can buff it with a towel. He will love that.

The step-by-step horse bath. First, groom your horse very thoroughly. This will loosen the dirt and move it toward the surface of the hair, where the shampoo will have a better chance of carrying it away. If your horse is shedding, remove as much hair as possible before you start the bath. The bath has seven steps:

1. Wet
2. Shampoo
3. Rinse
4. Condition
5. Rinse (if necessary)
6. Dry
7. Finish

1 Wet the horse. First, get your horse used to the feel and temperature of the water by slowly wetting his legs. Start from the hoof and work your way up. Then wet his entire body, one section at a time. For this you can use a hose (if your horse is used to it) or a bucket and a sponge. Here's a way to divide the horse into sections for wetting, washing, and rinsing:

- left side: shoulder from withers to knee; left side and back; left hindquarters to hock
- right side: (see above)
- tail
- left side of neck and mane
- right side of neck
- head
- legs

2 **Shampoo the horse** one section at a time. It would probably be safest to start with the near shoulder so you can get your horse used to the idea before you tackle the more sensitive areas like the legs, head, and tail. Squirt some shampoo solution onto a section of your horse. Scrub, add more water, scrub.

When you are shampooing, be sure to cover all sections of your horse. Don't skip these easy-to-forget areas:

- the throat area between your horse's jaw bones
- the area between your horse's front legs and just behind his elbows
- the area between his hind legs
- his belly
- his anus
- his sheath or her udder

You might need help with some of these areas because unless you are very experienced and have a gentle horse, you might have trouble getting the job done.

3 **Rinse the shampoo out of your horse's coat** until no more soap suds appear and his coat feels squeaky clean. Be sure to rinse all sections of your horse thoroughly. Start on the near side and rinse the neck, shoulder, ribs, back, and hindquarters. Move to the off side and follow the same pattern.

If his hair is still slippery, he needs more rinsing. Use a sweat scraper to remove excess water.

Wet your horse's legs to get him used to the feel of the water, and work your way up.

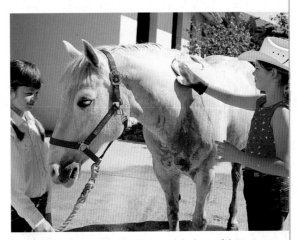

Apply the shampoo solution and lather with a rubber scrubber.

After shampooing and rinsing very thoroughly, remove excess water with a sweat scraper.

washing mane and tail

Don't just wash the hair of the mane. Scrub the crest of the neck, the inch-wide strip that runs along the top of the horse's neck where the mane hair grows from. Shampoo the crest really well and rinse it even better!

The same goes for the dock of the tail. The dock is a long and thick, fleshy covering over your horse's tail bones at the top of your horse's tail. This is where all of the tail hairs grow from. Don't forget to clean the underside of the dock.

The crest of the mane and the dock of the tail are what itch when a horse rubs bald spots in his tail or mane. You don't want that to happen.

4 **Condition the horse's coat** after all of the sections of the horse have been shampooed and rinsed. Pour the conditioning solution all over your horse.

Dilute your horse's coat conditioner in the following manner. Fill a small pail (one to two gallons) with water. Add a few good squirts of conditioner to the pail and mix it in well. Pour the conditioner solution slowly over the horse's mane, hindquarters, and tail; you might need to mix up several buckets to finish your horse. Coat conditioners can contain substances that smooth the hair, moisturize it, add shine, and protect the coat from the sun.

5 **Rinse out the conditioner if necessary.** Read the label of the conditioner carefully. If it says to rinse the conditioner out, then do so after about five minutes. If it's the type of conditioner to be left on, then of course you can skip this step.

6 **Dry the horse** using a sweat scraper (or your hands) to remove all water from the heavily muscled areas of the shoulder, back, belly, and hindquarters. The more you remove with the sweat scraper, the faster your horse will dry. Buff his head and legs with dry towels. Unless it is very warm and sunny, put a wool cooler on him so that he won't be chilled as he is drying. You can let him stand tied somewhere as he dries or you can lead him. If you turn him loose in a pen or pasture, the first thing he will do is roll, so keep him somewhere where he will stay clean until he is thoroughly dry and finished.

7 **Finish the horse.** Make sure all of the mane is lying on one side of your horse's neck. If it's not, place it there carefully with your fingers. Don't use a brush on the mane until it is completely dry, and use a comb only *after* you have thoroughly brushed the dry mane. The same goes for the tail — wait until it is completely dry. You will damage and break many tail hairs if you try to comb through it when it's wet. If you have a finishing spray for the mane and tail, apply it when the tail is almost dry and work it in by drawing your fingers through the mane and tail.

Clipping

If your horse is not perfectly mannered about clipping, you will need someone to train him for you before you try to clip him yourself. He will need to get used to the sight, sound, and smell of the clippers and the feel of the vibration of the clippers on his body. In the meantime, you can use scissors to do some of the trimming. It's best to wait until your horse has been bathed and is completely dry before clipping; clipping dirty hair will make your clipper blades dull.

When you are clipping, be sure the electric cord is out of the way so neither you nor the horse becomes tangled in it. The clippers must have sharp blades or they will pull on the horse's hair and cause him to dislike being clipped. Even if your horse is well trained about clipping, let him see and smell the clippers before you begin. Rub the clippers on his neck with the motor turned off, then turn them on. Turn them off and move the clippers to the bridle path area or to his legs, wherever you want to start clipping. Leave the clippers off and move them around the area as if you are clipping, then turn the motor on.

Don't clip the hairs around the nose and eyes and inside the ears because these hairs protect your horse. Nature has provided face whiskers so your horse can put his head into a space that he can't see without getting hurt; the whiskers act like antennae to tell him when he is getting close to bumping into something. If you clip these whiskers off, he will lose his "second set of eyes." If you remove the hair from inside his ears, tiny gnats and flies will be in for a feast, which will be very painful for your horse and could make him head shy. So, when clipping, just tidy up the bridle path and legs.

clipping the bridle path and fetlocks

How long should the bridle path be? I suggest clipping a two- to three-inch bridle path, which will give the halter and bridle a neat place to lie. Clip all the long fetlock hairs, as this will help keep your horse's legs clean.

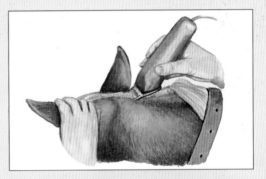

The bridle path should ideally be 2–3" long.

Clip the fetlock hairs so they don't pick up dust and mud.

Blanketing

If you will be keeping your horse in a stall a good deal of the time, you may need to invest in some stable clothing for him. Blanketing keeps your horse cleaner, so that grooming takes much less time. A blanket that is too heavy is unhealthy, however, because it may make the horse sweat and then become chilled.

A blanket must fit a horse properly or it can cause rub marks and sore spots on his withers, shoulder, chest, and hips. A blanket that is too large will slip and twist and can even slip so much that the horse could become dangerously tangled. To be sure you buy the correct size blanket, measure your horse from the center of his chest along one side of his body to the center of his tail.

The blanket lining must be of a smooth material to prevent damage to hair, especially the mane area nearest the withers. Another problem you should be on the lookout for is overheating. If your horse is wearing a very warm winter blanket and you have a sunny, calm winter day, your horse can become dangerously hot. Also, waterproof blankets that aren't breathable don't allow the horse's body heat and sweat to escape. If your horse overheats and sweats during the day, he will be cold and chilly at night, and he could get sick. When you are having warm weather in the middle of winter, check to see if your horse is overheating by slipping your hand under his blanket at the heart girth area.

A horse should have his blanket taken off and shaken out every day. Hang the blanket inside out in the sun while you are riding or exercising the horse. Then cool him out thoroughly, brush him, and put the blanket back on.

Keep your horse's blanket clean.

A dirty blanket will cause your horse discomfort and may even cause disease. If you keep the horse clean, the blanket will usually stay clean on the inside. But if you let mud and manure build up on the outside of the blanket, the extra weight may be uncomfortable for your horse and may cause the blanket to shift to one side. Also, dirt and manure can rot the material.

Clean the outside of the blanket by brushing, shaking, and even hosing it regularly. It will also be necessary to wash the blanket from time to time. If your horse lives outside and rolls on the ground, you might need to wash his blanket every two or three weeks. If your horse and his stall are kept clean, perhaps you will need to wash the blanket only once or twice a year.

blanket care

Check the blanket regularly for loose stitching or straps that are frayed or cracked. Have the necessary repairs made before the tears become so big that the blanket is ruined and no longer worth mending.

types of blankets

Cooler. Usually made of wool, so you will need to hand wash in cold water. Coolers are usually draped loosely over a horse after a bath or exercise while he is cooling out or drying off.

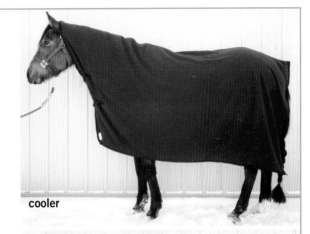
cooler

Sheet. Usually made of cotton. These will shrink, so if you wash your horse's blanket in hot water you will have to give the sheet to your little sister's pony! Cotton sheets are light protection from dust, flies, and drafts and are used as a protective cover during late spring, summer, or early fall.

Fly sheet. A fine mesh sheet with large holes. Used in summer so the horse doesn't get hot but flies stay off his body.

sheet

Winter blanket. Many types are available, and most are designed to be used in a barn because they are not weatherproof. They might be made of canvas with a wool lining or quilted just like the jacket you wear for skiing. They are machine washable in a large machine at the laundromat. Winter blankets usually have two belly straps, a chest strap, and sometimes hind leg straps. You can usually get a matching hood that covers your horse's head and neck.

Turnout rug or blanket (often called a New Zealand rug). This type of blanket might be made of waterproof heavy canvas with wool lining or other waterproof, breathable materials. A turnout blanket needs to be rugged and tough because it is worn by horses that live outside for the winter.

turnout blanket

chapter six

Shelter and Pasture

A horse's shelter requirements are pretty basic. In the summer be sure there is a place for your horse to get shade — in a shed, in a barn, or under a tree. When flies are bad, help him ward them off with a fly sheet, a mask, and the careful use of fly spray (see Chapters 5 and 7 for more information).

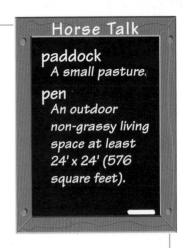

Horse Talk

paddock
A small pasture.

pen
An outdoor non-grassy living space at least 24' x 24' (576 square feet).

In the winter, when temperatures drop below 50°F, be careful your horse doesn't get chilled. If it's raining or snowing and there is a wind, your horse should have shelter. An alternative is to have your horse wear a turnout blanket (see Chapter 5).

Your horse can't go get a glass of water when he is thirsty or put on a jacket when he gets cold. He depends on you. Be sure he has what he needs for good health and comfort. Whether you are keeping your horse at home or at a boarding stable, his welfare is still your responsibility.

How you take care of your horse depends on where he lives — on pasture or in a *pen, paddock,* or stall.

Keeping Your Horse on Pasture

This is the most natural way to keep your horse, but it has disadvantages as well as advantages.

A big plus of pasture living is that your horse can have free exercise, fresh air, and sunshine. If there are other horses there, pasture living allows your horse to socialize. Depending on the quality of the pasture, a horse might get all or just a very small part of his feed needs met. Many horses stay more mentally content on pasture than if they are kept in pens or stalls. Also, you will probably have fewer daily chores if your horse is on pasture, because you won't have a stall to clean.

The reason not all horses can be kept on pasture is that land is expensive and horses are very hard on land. In some areas, such as Iowa and other Midwestern states, pastures are so lush that one acre provides plenty of grass for one horse. In other areas, such as the dry ranges of Wyoming or the desert of New Mexico, pastures are very dry and thin, and it might take more than fifty acres to provide enough grass for one horse. You will need to talk with your county extension agent to find out how many horses your pasture can support. (See Chapter 8 to learn how to find your county agent.)

Contact your county extension agent to learn the "carrying capacity" of your pasture, and never put more horses on it

Keeping horses on pasture is natural and allows free exercise, fresh air, and socialization. But it can also make horses "herd-bound."

than that. If you have two or more pastures, rotate your horses between the pastures and the holding pens so the grasses have time to regrow.

Another disadvantage of keeping a horse on pasture is that he might become so closely bonded to other horses in the herd that he would be hard to catch and would resist being taken away from the other horses to go for a ride.

Some horses are very grouchy. When put in a pasture with other horses, they might kick and fight. If your horse is grouchy or has to share a pasture with a grouchy horse, he could be injured.

If there is more than one horse on the pasture, it is difficult to be sure that each one is getting the appropriate amount of hay or grain for feed.

It usually takes more time to catch and groom a horse kept on pasture, which means it takes longer to get ready for riding.

Flies are more of a problem for horses on pasture because damp, grassy areas provide good breeding grounds for flies.

The most serious disadvantage to keeping a horse on pasture is probably that, like most horses, he may not know when to quit eating and can get very fat. When a horse eats too much, he can get a very bad stomach ache, called colic, or he could founder, which is a serious lameness affecting the hooves. Both conditions are described in Chapter 7.

Make sure your fences are safe so your horse doesn't get hurt. There are many types of safe horse fence, such as wooden or synthetic board, post and rail, pipe, and electric. Whatever type you choose, be sure it is at least five feet high; six feet is even better. Check the fence routinely for damage.

Horses like to hang around gates, waiting to be fed or wanting to be groomed and ridden. These "gate potatoes" often crowd the gate area or press over the top of the gate. And when play gets rough, they might push right through the gate. That's why the gate must be at least as tall as the fence, must be made of safe, strong material, and must be securely latched with a horseproof latch. Horses are pretty quick learners when it comes to opening gates.

If your horse gets loose, whether he jumps the fence or walks out through a gate that was left open, you are responsible for any trouble he might get into. If he ruins your neighbor's prize garden or walks out on the road and causes a car accident, you are responsible. That's why it is important to keep all gates securely latched and to maintain fence safety.

One way to be sure your fences are in good condition is to make fence checking a fun part of your daily routine. Saddle up your favorite horse and "ride fence," making note of anything that needs to be fixed. Later you can go back with the necessary tools and materials to repair the fence.

fence safety tips

★ **Replace barbed wire with safe horse fencing.** Barbed wire is not a safe type of fence for horses. It can cause serious wounds that could result in permanent lameness.

★ **Repair broken boards.**

★ **Remove and replace nails that are sticking out.**

★ **Replace fence that is lower than four feet tall** (unless you have only small ponies).

★ **Immediately treat any wood that shows signs of being chewed.**

Keeping Your Horse in a Pen or Paddock

A pen or paddock is often a good compromise between keeping your horse in a stall or keeping him in a pasture. A pen is an outdoor living space for your horse that is at least 24' x 24' (this would be the size of about four stalls put together.) A pen does not have grass growing in it, but a paddock does — it is a small pasture anywhere from about a quarter acre to several acres in size. Besides providing exercise, a paddock usually also provides some grazing. A run is a long, narrow, fenced-in area that is usually attached to a stall. A common-sized run is about 12' x 60', which allows a horse to trot a few strides, turn around, and come back.

If the pen, run, or paddock is large enough and your horse exercises by himself, then you might not have to take him out of the pen for extra exercise every day. But even if your horse's pen is fairly large, you still should take him out for riding or longeing at least three or four times a week.

CAUTION Metal farm panels are often used to make temporary, movable pens. Some panels are very dangerous for horses. For instance, some have places where the horse can get his legs trapped when he rolls near the bottom of the panel. If the panels are shorter than five feet, the horse could get his legs over the top if he rears in his pen. Also, with some panels, a horse can get its leg caught between the panels.

A paddock for winter turnout must be safe because the footing is often slippery and the horses are frisky!

If you keep your horse in a stall, you'll need to provide him with daily exercise, which can include turnout, riding, longeing, driving, or in-hand work. Always keep safety in mind when handling your horse.

Keeping Your Horse in a Stall

Horses that live indoors are healthiest and happiest when the building has lots of fresh air but is draft free and dry. An airtight, heated barn is usually damp and very unhealthy for horses.

A 12' x 10' box stall works well for most horses; larger horses (over 16 hands) will require a 12' x 12' or 12' x 14' stall. Arabians are often comfortable in a 10' x 10' stall. Ponies can get along in stalls 8' x 10' or smaller, depending on the size of the pony.

The stall floor should be a material that drains well, such as a mixture of very small gravel and dirt. If you add mats or rubber stall flooring over the soil, they will keep your horse cleaner and drier, and the stall won't develop huge craters from your horse pawing, rolling, and lying down.

The type of bedding you use will depend on what is available in your area. Wood shavings, straw, and peanut or rice hulls all make good horse bedding. Try the types that are available where you live and compare how much they cost, how long they last, and how comfortable they are for your horse. You will still need to use bedding with flooring, but you will find you use much less.

When a horse lives in a stall, be sure he gets exercise every day. The best kind of exercise for a stalled horse is free turnout followed by at least one hour of continuous, controlled exercise such as riding, longeing, or driving. If you just turn your horse out for exercise, often he will take one gallop to his favorite corner of the pasture and eat. Or, if a horse has not been out for a few days and you turn him loose, he may run and play so hard to burn off extra energy that he injures himself.

Horses in stalls are right there when you want to ride and are usually clean enough to need only a minimal grooming. If you blanket a stalled horse and keep the barn lights turned on for a few hours before dawn and a few hours after sunset, he will tend to grow less winter hair. This makes it easier for grooming and also will let him cool out more quickly after exercise. But once you blanket a horse in the fall, you will probably have to keep him blanketed all winter. It wouldn't be fair to turn him out in the cold weather without a blanket once he is used to it. See Chapter 5 for more on blankets.

It is easier to clean a stall (or pen) when the horse is turned out for exercise. When it is time to put your horse back into the stall or pen in the evening, rake the dry "old" bedding back to the area where your horse usually defecates and urinates. Add some new bedding, if necessary, to the place where your horse likes to lie down.

Every week or so, you will need to remove all of the bedding from the stall. This is called "stripping the stall." Your horse will appreciate a nice clean stall, and he'll be healthier, too.

Before you return your horse to his stall, pick out his hooves and give him a good brushing. You don't want him to bring mud or manure into his freshly cleaned stall.

cleaning a stall or pen

To clean your horse's living quarters, use a fork, shovel, and freshener.

Remove the manure using a special fork that has tines close together.

Search for the spots of wet bedding and remove them. Sweep the dry bedding against the stall walls so the wet spots on the floor can dry.

If you like, you can sprinkle a barn freshener or lime on the wet spots. Let the stall floor dry all day, if possible, with the barn doors and windows open.

three ways your horse sleeps

Horses rest in three ways: **standing, sternal recumbent,** and **lateral recumbent.** It is perfectly natural for a healthy, fit, sound horse to lie down, but if your horse spends a large amount of time lying down, he may have hoof or leg pain. Also, lying down for long periods of time can cause digestive problems.

1 Standing. The most common way a horse rests is dozing while standing. Horses have a specialized leg design that allows them to sleep while standing without falling down! A horse can get quality rest while dozing on his feet if he has a level place to stand that is protected from bad weather, loud noises, bright lights, and insects.

2 Sternal recumbent. When a horse lies down on a sunny day or after a hard workout, he'll start in the sternal recumbent position by kneeling, tucking his hind legs under his body, and lying on his belly, sometimes resting his chin on the ground. Horses sleep soundly in this position but, if startled, can rise in an instant. Provide a soft place for your horse to lie down so he doesn't develop sores on his knees and hocks when he gets up.

3 Lateral recumbent. When a horse is very relaxed, he might roll over on his side and sleep in the lateral recumbent position, flat on one side, with all four legs extended and head and neck on the ground. Because a horse straightens his legs in this position, his legs can get tangled in a fence or trapped against a stall wall, so make your facilities safe and check your horse often. It takes more time for a horse to get up from the lateral recumbent position but it can still be quick, so be careful when approaching your sleeping horse.

chapter seven

Health Care

A healthy horse is a happy horse. The best way to keep your horse healthy is to design a daily routine and then stick to it. This routine should include a schedule of feeding and exercise plus a daily health check. The daily health check is the best way to catch a problem before it becomes serious. Learn all you can about health care — your horse depends on you.

The Daily Routine

In the morning, feed your horse his hay and give him a quick health check (described below). Feeding his hay first will take the edge off his appetite and keep him from gulping his grain. Be sure he has enough water and that it's clean. Then feed the grain ration.

Sometime after your horse has finished his morning ration, take him out of the stall or pen and tie him in a grooming area. As you pick out his hooves and groom him, give him a closer check. Then either turn him out for his daily exercise or ride him. Clean the stall or pen and then re-bed it. Return the horse to his stall or pen. At the evening feeding, again, feed hay first, check the water, and then feed grain.

If your daily chores involve pasture horses, you will have a different routine. Just because pasture horses might be farther away and you don't see them as often, don't forget them. Be sure you see them every day, and give each one a thorough check.

Every day you should give your horse a visual health check to get an idea of how he is feeling. If you spot something suspicious, you'll need to perform some hands-on tests. First, however, you need to learn the signs of a healthy horse. Then you need to learn the signs of an unhealthy horse. Once you know what to look for, you should be able to make a daily check in a couple of minutes. And when you know what's normal for horses in general, then you can figure out what's normal for your horse. Just like people, every horse is different.

➲ How is your horse standing? Is his head down or up? If it's down, he might just be dozing, or he might be feeling sick. Is he holding one leg up? If it's a hind leg, he might be resting it while he is sleeping, or it might be lame. If it is a front leg, it probably is lame.

➲ How is he lying down? Is he in a normal, peaceful sleeping position? Or is he restless and rolling back and forth?

What is his expression? Is he alert with ears forward and eyes bright? Does he look content? Or does he look dull or nervous?

Check his legs. Look for wounds and for swelling or puffiness. If you notice something out of the ordinary, you should halter your horse and examine his legs by feeling them. You'll need to develop a feel for what is normal texture and temperature for horses' legs in general and what is normal for *your* horse's legs. When your horse moves, does he place weight on all four legs equally or does he limp? Does he bob his head, skip, or buckle over at the hoof? Does he take short, stiff steps? All of these can indicate a lameness problem.

Check his appetite and thirst. Has your horse finished all of his feed from the previous feeding? Has he drunk plenty of water? Is he standing by his feeder at feeding time waiting for his next meal? A good appetite is one of the signs of good health. Your horse should finish all his food within two to three hours.

Check your horse's manure. The fecal balls should be well formed but should break in half easily. If the fecal balls are dry and hard, the horse is not drinking enough water. Loose sloppy piles (more like "cow pies") tell you that your horse's feed is too rich (too much grain or alfalfa hay), he is getting too much salt and water, or he has an irritation in his digestive tract and has diarrhea. If you see slime or mucus on his manure, he has an irritated gut. If there is whole grain in the manure or long pieces of fiber from hay, it means the horse is gobbling his feed without chewing, the feed is passing through his

Normal manure consists of well-formed balls with enough moisture that the pile stays heaped. It is normal for a small amount of hay to pass through undigested.

body too fast, or he has a dental problem and can't chew his feed thoroughly. If you see worms in his manure, this tells you that it is past time for you to deworm him.

⊘ **Notice anything unusual.** Examine his stall or pen and his body for signs of rubbing, rolling, or pawing. Is his tail ruffled? Is he covered with dirt or manure? Is he sweaty? Are there holes in his stall from pawing?

The Vital Signs

If you have reason to suspect there might be a problem such as colic, you should check your horse's vital signs. Your veterinarian can show you how to do these simple checks.

⊘ **Pulse.** Pulse rates can be taken at the horse's jaw or just above the fetlock. The average pulse rate of an adult horse at rest is about 35 to 40 beats per minute. Young horses have much higher pulse rates. A two-week-old foal's pulse can be as high as 100 beats, and a two-year-old's can be 40 to 50 beats per minute. If your horse is excited, in pain, or nervous, has a high temperature, is in shock, has a disease, or has just completed exercise, his pulse rate will be higher than normal.

⊘ **Respiration.** The average respiration rate of an adult horse at rest is 12 to 25 breaths per minute. One breath in plus one breath out equals one breath.

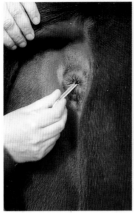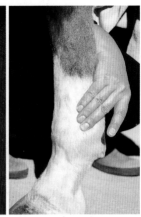

LEFT: Apply a dab of lubricating jelly to the thermometer. Hold the tail to one side and ease the thermometer into the anus, in a slightly upward direction, until about 2 inches remains outside.

RIGHT: You can take a horse's pulse rate at the digital artery, on both the inside and outside of the leg just above the fetlock.

To measure your horse's breathing rate, watch his flank area and count every time he breathes in as one breath. This will be easier to see after your horse has exercised. It takes a lot of practice to get an accurate respiration count when a horse is resting.

⊘ **Temperature.** The average temperature of an adult horse at rest is about 100°F; normal is usually about 99 to 101°F. Use a farm animal thermometer and thread a string through the eye. Attach a clip or clothespin to the end of the string, and clip this to your horse's tail. Your 4-H leader or veterinarian can teach you to shake down the thermometer, then lubricate it with petroleum jelly or a drop of saliva and insert it into the horse's rectum. It should be left there for two to three minutes before you take a reading.

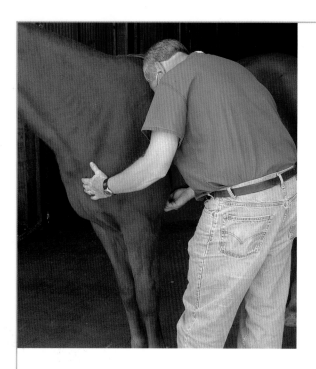

Preventive Health Care

"An ounce of prevention beats a pound of cure." This old saying couldn't be more appropriate than for horse health. If you take the time to do the small things, you will avoid big problems. To keep your horse from suffering fatal conditions such as colic, founder, or disease, you just need to pay attention to simple details like regular feeding, deworming, vaccination, and keeping your horse's living areas clean.

Parasites are a common health problem for horses. You should learn about the kinds of *parasites* that might threaten your horse's health and how to protect your horse from them.

Internal: worms and bots. All horses have worms, which are also called internal parasites, meaning that the parasites live inside the horse's body. If worms are allowed to breed uncontrolled inside your horse, pretty soon there will be so many worms that they eat more of your horse's feed than he does.

All horse manure contains worm eggs. After the horse drops the manure on the ground, the worm eggs hatch. The tiny little worm larvae are so small you can't see them easily. They crawl up on blades of grass and are eaten by the horse. Once they are inside the horse, they become big worms, which live in his stomach and intestines and eat his hay and grain. They lay eggs and continue the cycle.

Horses also have trouble with bots. Bots are similar to worms because they live inside the horse like worms do. But instead of hatching into larvae that crawl onto grass, bot pupae turn into botflies. These flies look like bees (but they don't have stingers) and buzz around your horse, laying eggs on the hair of his legs and body. The tiny yellow eggs appear in

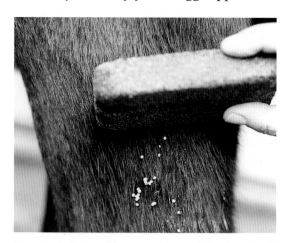

Remove bot eggs from your horse with a bot block.

clusters that are easy to recognize. When your horse scratches his leg with his mouth, he swallows the eggs and the whole cycle starts all over again.

When botflies are buzzing your horse, it can be very dangerous for you to handle him. One type of botfly tries to fly up the horse's nose, which drives most horses crazy. They may strike at the flies with their front legs and run frantically to get away from them. Be careful in late summer when the botflies are around your horse. As soon as you see bot eggs in late summer or fall, remove them by scraping with a **bot block** (available at your farm supply store or tack shop) or a dull pocket knife.

It is probably easy for you to see why you need to make a parasite prevention plan and stick to it. Most horses need to be dewormed every two months. That's six times a year. Ask your veterinarian to help you make a plan and choose the dewormers you should use. If you watch your veterinarian carefully, you can learn how to deworm your own horse.

External: ticks and lice. Pests that live on the outside of horses' bodies are called external parasites. Lice and ticks often live in a horse's coat. They like to burrow in the mane and tail and can make your horse itch like crazy. In certain parts of North America, ticks also carry Lyme Disease, which is contagious to humans. If your horse rubs bald spots in his mane or tail, have your veterinarian check him for ticks or lice and recommend a treatment program.

worm cycle

The life cycle of most internal parasites starts when the horse eats larvae with his pasture or hay. Once inside the horse, the larvae develop into adult worms and lay eggs. The eggs are dropped on the ground with the horse's manure and hatch into larvae, which the horse eats, starting the cycle all over again.

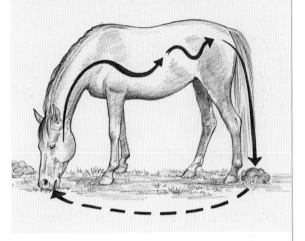

Fungus and other skin problems.
Itching can also be caused by tiny organisms that grow on your horse's skin. Ringworm will make your horse's hair fall out in a patch the shape of a circle; that's how it got the name ringworm, even though it's not a worm at all.

Ringworm is contagious — you and your other horses can catch it — so if you have a horse with ringworm, take care of it immediately before it spreads. Otherwise, there will be a fungus among us! Clean the area with medicated soap and apply the ointment recommended by your veterinarian. Keep all grooming tools that you have used on the infected horse separate from other items. Wash them in the disinfectant recommended by your veterinarian. Halters, blankets, stall walls, and fence posts that your horse has rubbed on will also carry the fungus, so disinfect them too.

Another itchy skin problem is rain rot, which is usually caused by a bacteria. If your horse has this, he will have crusty yellow scabs and bumps on his neck, back, and actually anywhere on his body.

You will need your veterinarian's help in figuring out what is causing your horse to itch. Once she determines what it is, she will give you the proper medicated shampoo and instructions on how to cure your horse's problems. (See page 54 for information on giving your horse a bath.)

Keep a record of your horse's health by using a little notebook or a card file (like a recipe box) to write down important

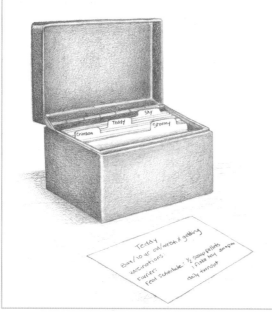

information about your horse's health. For each horse, you should keep complete records on health care, farrier work, training, and breeding. Use a calendar to remind yourself of deworming schedules or farrier visits.

Besides keeping daily records, keep a file folder of important documents in a safe place in your house. Items like the horse's registration papers, the bill of sale you received when you bought the horse, a brand inspection certificate, a registration of his tattoo or freeze branding, his pedigree, his insurance policy, and any important test results such as the Coggins test for equine infectious anemia all belong in the folder. Include at least four clear photos of your horse showing him from both sides, the front and the rear. That way, if your horse ever wanders down the road or is stolen, you will have a way to identify him to the police.

Vaccinate at least once a year to protect your horse against getting certain diseases. Your veterinarian will tell you what vaccinations your horse should receive, depending on where you live. Almost all horses should be vaccinated against tetanus, encephalomyelitis, influenza, and rhinopneumonitis. Often one vaccination will cover all of these diseases. Such a vaccine is called a "four-way" or "five-way" vaccine. In some parts of the country, your horse will also need to be vaccinated for Potomac horse fever, rabies, and strangles.

To help prevent disease and infection, keep facilities and equipment clean. Wash your horse's water pails and feed buckets regularly. Discourage flies and rodents by raking up loose hay daily around the barn.

Ways to Prevent Disease and Infection

Disease and infection are spread in several ways: directly from one horse to another horse, from a contaminated stall or feeder to a horse, between horses eating or drinking from the same place, or through the air.

If you think there is a contagious disease at your place or at your neighbor's, or if you have been to a stable or horse show where there were sick horses, you will need to work closely with your veterinarian to keep disease under control. Use a combination of treatment, disinfection, and quarantine to prevent the disease from spreading and to get rid of the bacteria or virus that caused it.

Treatment usually means drugs that your veterinarian will give your horse. Disinfecting will require you to use a special medicated soap to scrub your horse's feeders, waterers, and living quarters to get rid of the microscopic "bugs" that caused the disease. Sunlight is a very powerful disinfectant as well, especially if the weather is hot and dry.

Whether you have a disease problem or not, any new horse should be quarantined and observed for at least a week before you allow him to come in contact with your other horses or their eating or drinking areas. Any horse that leaves your place should be quarantined upon return, especially if it has been exposed to a large number of other horses, such as at a horse show.

common diseases

Disease	Description	How It's Contracted	How It's Prevented
Tetanus Common name: lockjaw	The horse's muscles stiffen up so severely that within a few days he dies or must be put to sleep.	The bacteria that cause this infection of the nervous system enter the horse's body through a wound or through the umbilical cord that is attached to a foal's belly button.	A yearly tetanus vaccination.
Equine Encephalo-myelitis Common name: sleeping sickness	The horse develops a high fever and then is paralyzed and dies within two to four days.	A mosquito transports the virus from a wild bird or animal to your horse.	A yearly vaccination. All horses in the U.S. and Canada should be vaccinated against Eastern (EEE) and Western (WEE) encephalomyelitis. There's also a Venezuelan strain. Ask your vet if you need to include VEE in your horse's vaccination program.
Influenza Common name: the flu	The horse will have a fever, a runny nose, and a cough. Many horses get the flu, but few die from it. The flu can be treated by your veterinarian.	This virus is carried in the air. When one infected horse coughs, another can breathe in the germs and get sick.	Vaccinate your horse once or twice a year, especially if there is a flu problem near where your horse lives or if you travel a lot with your horse.
Rhinopneu-monitis Common name: a cold	The horse has a lot of white discharge coming out of his nose.	It usually affects foals that are four to six months old when they are weaned from their mothers. If a pregnant broodmare comes in contact with this virus, she could lose her fetus (abort).	Quarantine horses that have Rhino and vaccinate once or twice a year against the disease. Check with your veterinarian as to how many times you should vaccinate; if your farm is at risk, the vet may encourage you to vaccinate twice or even more per year.

Disease	Description	How It's Contracted	How It's Prevented
Distemper Common name: strangles	The horse will not eat or drink and gets a very high fever. A lot of thick, yellow pus comes out of the horse's nose and the ruptured area near the throat. Occasionally, a horse will die from strangles.	A bacterial infection causes the glands near the throat to swell and eventually rupture.	You will need to disinfect everything an infected horse comes in contact with because the bacteria are very contagious. Ask your veterinarian whether you should vaccinate against strangles.
Rabies	This disease rarely affects horses, but when it does, it usually results in violent, dangerous behavior and death.	The rabies virus is transmitted from an infected animal to the horse by way of a bite, usually from a dog, a skunk, a fox, or a bat.	If rabies has been reported in your area, your veterinarian might suggest that you vaccinate your horse against the disease.
Equine infectious anemia Common name: swamp fever	The horse might not show symptoms and just be a carrier or he might show fever, depression, weight loss, and swelling. This is a potentially deadly disease.	This virus lives in the horse's blood and is spread from one horse to another through a biting insect.	There is no vaccine to protect your horse against swamp fever but there is a test, called the Coggins test, to see if your horse has been exposed to the disease and is a carrier. A Coggins test is required if you are going to travel out of state with your horse or take him to a horse show or clinic. If a horse is found to be a carrier of EIA, depending on what state you live in, he might have to be put to sleep.

Ways to Prevent Poisoning

Horses like to investigate unknown things by nibbling and tasting. To be sure your horse cannot poison himself, keep all dangerous substances out of his reach.

⊙ **Paint.** Be sure the paint that is on your horse's fence and pen rails, and on any building near him, is safe. Paint that contains lead can be very harmful to your horse if he swallows it.

⊙ **Horse products.** Before using *any* product on or near your horse, read the label carefully. Make sure you don't accidentally give your horse an overdose of an antibiotic, a dewormer, or a nutritional supplement. Some dewormers (organophosphates) can be poisonous when they are used several times in a row because they can store up in a horse's body. That's why you should be sure you know what you are using when deworming.

⊙ **Grain and seeds.** Don't feed your horse grain or seeds that are meant for planting because they might have been treated with mercury, which will harm your horse. Treated grains often look pink or reddish, but sometimes they look exactly like the grain you normally feed your horse. So be extra sure!

Don't give horses feed that was meant for cattle, sheep, or goats. These feeds often contain urea, a type of feed additive that is okay for ruminant animals (those with four-part stomachs) but not for horses, whose stomachs are very different.

Also, some cattle feeds may contain growth stimulants that can permanently damage the nervous system of a horse.

⊙ **Toxic materials.** Don't let horses get near junk or vehicles. Because horses use their lips to inspect things, they might eat poisonous paints, antifreeze, or battery fluid. Protect your horses from toxic fumes from vehicles, paints, and solvents. Don't apply insecticides or herbicides near their feed or water areas, and be aware of which way the wind is blowing when you are spraying.

Dental Care

Your horse should have a dental checkup at least once a year. If your horse is more than five years of age, probably all he will need is a "floating." When the veterinarian or equine dentist floats your horse's teeth, he uses a large file and smoothes the sharp edges of your horse's molars.

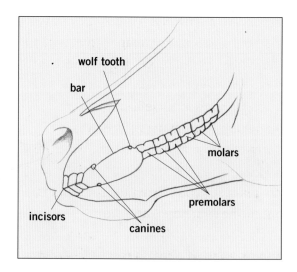

A horse has five different types of teeth.

Young horses usually need their teeth floated every year, too. They also need two other checkups. Sometimes two- and three-year-old horses don't shed their baby teeth properly, so the baby teeth get stuck on top of the adult teeth that are trying to push up. This makes it painful and difficult for your horse to chew. Your veterinarian can spot this problem and easily pop the baby "caps" off the adult teeth.

Some horses have a small tooth, called a "wolf tooth," in front of their molars. This tooth can cause problems if you use a snaffle bit on your horse, because your horse's lip can get caught between the bit and the wolf tooth. You should have your horse checked for wolf teeth when he is about one year old; if wolf teeth are present, your veterinarian can remove them.

Hoof Care

Hooves grow about one-quarter inch per month and, like your fingernails, must be trimmed regularly. Otherwise, they grow too long and break, damaging the hoof. Most horses need professional hoof care every six to eight weeks.

Be sure your horse has very good manners while having his feet handled and worked on so your farrier doesn't get hurt. Make sure you have a clean, level, dry place for your farrier to work that is out of the sun, wind, and rain

Moisture control. It is not good for a horse's hooves to be too wet or too greasy or oily. Mud and water cause the hooves

daily hoof check

Every day before and after riding, you need to check your horse's hooves for rocks, splinters, loose shoes, and loose nails. While you are checking, this is a perfect time to pick the hooves clean.

to spread out like a pancake and split and break. Do not overflow your horse's water trough and make him stand in the mud. Do not slather on hoof dressing. Hooves are healthiest when they are kept clean, dry, and trimmed regularly.

A liquid hoof sealer can help your horse's hooves by sealing out external moisture and sealing in moisture from the blood supply. Hoof sealer is a thin, clear liquid that you paint on the hoof. It is not hoof dressing, which is a greasy salve sometimes used on dry hooves.

parts of a hoof

Know the parts of the hoof so you can talk with your farrier.

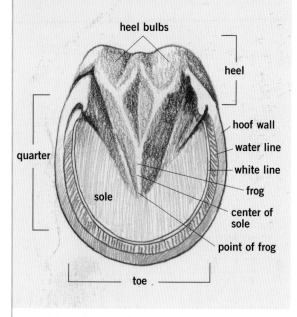

heel bulbs

heel

hoof wall

water line

white line

frog

center of sole

point of frog

quarter

sole

toe

Trimming. Trimming should be done by an experienced horseshoer. If your horse stays barefoot, he should receive a trim at least every six to eight weeks.

Shoeing. If you are riding your horse a lot, he might need shoes to protect his hooves from too much wear. If your horse's hooves are crumbling or have cracks, he might need shoes to help hold his hoof walls together whether or not you are riding him. If your horse is wearing shoes, they should be checked by you daily and by your horseshoer every six to eight weeks.

Cleanliness. Manure that becomes packed in a hoof makes a perfect home for bacteria and fungus that can cause the hoof to crumble and fall apart. (See discussion on thrush on page 89.)

Exercise. Exercise is important for healthy hooves because when blood flows around your horse's body, his legs and hooves get their share. Fresh blood carries oxygen to the tissues of your horse's hooves and nourishes the hooves so they can stay healthy.

If your horse is forced to stand still in a stall or a small pen, the circulation to his legs is very slow, and the blood in his legs and hooves is "old, tired" blood.

Sanitation

Sanitation is a big word for keeping things clean, clean, clean! The cleaner you keep your horse and his stall, pen, or pasture, the healthier he will be. Sanitation includes cleaning up manure, keeping the area around your horse's living quarters dry, and keeping flies and other pests to a minimum.

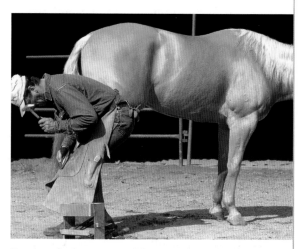

Be sure your horse has very good manners for your farrier's visit.

Manure. Whether your horse lives in a pasture or a stable, he will produce up to fifty pounds of manure every day! If manure and urine build up for just a few days in an enclosed barn, they produce a smell and gas (ammonia) that will burn your eyes and lungs . . . and your horse's! Also, if your horse is forced to stand in wet manure and urine, it will cause his hooves to break down. It can also cause him to suffer from a painful hoof condition called thrush (see page 89).

Since all manure contains worm eggs, you need to pick it up every day and get it away from your horse. Once you pick up the manure you have three choices:

⟳ You can have the manure hauled away. Some garbage collection services will carry manure away with your garbage.

⟳ You can spread the manure on a field or pasture that won't be used by horses during the year. It is best if it is spread very thin and exposed to the sun and drying wind, which will kill the worm and fly eggs.

⟳ You can store the manure in a pile to spread later. When manure is stacked in a neat pile, it will compost or break down and turn into humus, which is a dark, fine substance with no odor. Humus is valuable as an addition to the soil of your garden or flower beds or for spreading on a pasture. It will take from two weeks to three months for a pile of manure to turn into humus, depending on your climate, rainfall, and the type of bedding that is

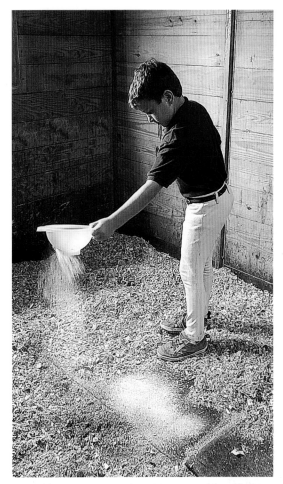

A dry stall will keep your horse's hooves healthy, minimize odors, and discourage flies from breeding.

mixed in with the manure. You will probably want to keep two manure piles going all the time — one is the old pile that is "cooking" or turning into humus, and the other is the fresh pile that you are adding to every day.

Flies. Stable flies, horseflies, deer flies, horn flies, and face flies are all bloodsuckers that can cause problems for both you and your horse. The most common type of fly that will bother your horse is the stable fly.

four ways to control flies

The best way of controlling flies is to prevent them from breeding in the first place.

1. Remove breeding grounds by picking up manure at least once a day. Keep your horse's living quarters dry.

2. Be sure all pens and stalls drain well.

3. Repair leaking faucets, hoses, and waterers.

4. Dry out wet spots in stalls and pens by clearing the wet bedding away, adding lime, and letting the ground dry.

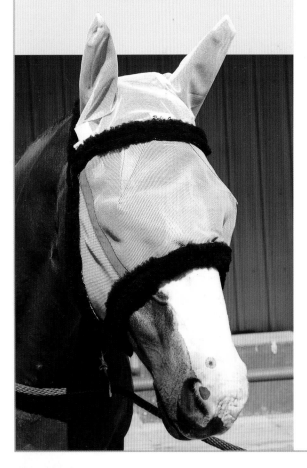

Stable flies are about the same size as the common housefly. However, stable flies suck blood while houseflies do not. Stable flies bite your horse until he bleeds and then feed on the blood. They like to feed on a horse's lower legs, flanks, belly, under the jaw, and at the junction of the neck and the chest — all the places where your horse's skin is the thinnest and easiest for a fly to bite through.

The bite from a stable fly is painful. Some horses panic when flies are after them and could be injured trying to get away from the flies. Even tough horses that try to put up with flies still stamp their legs to keep the flies from biting. This is very hard on their legs and joints and can cause the horse's shoes to become loose.

Stable flies lay their eggs in manure, wet hay, unclipped grassy areas, and other places where there is moist plant material. A female stable fly might lay twenty batches of eggs during her thirty-day life span. Each batch contains between forty and eighty eggs. It takes twenty-one to twenty-five days for the eggs to hatch. When the eggs hatch, the adult flies emerge ready to breed. (If you have seen tiny flies on manure and thought they were young stable flies, you were probably looking at a different type of fly.) The number of flies that can be produced by just one female is 1,600! If each of her 1,600 kids had 1,600 kids, that would make two million five hundred and sixty thousand (2,560,000) flies! And that's just from one female fly to begin with!

Even if you are the best horsekeeper in your state, you still will have a few flies to deal with. But you will have to use fewer fly sprays if you keep your facilities clean and dry. If you carelessly use insecticides (chemicals that kill flies) or repellents (chemicals that keep flies from landing on your horse), you can cause harm to your horse, yourself, and the environment.

If necessary, you can use spray-on or wipe-on fly repellents when the flies are the worst. Ask your 4-H or Pony Club leader to help you decide what product to use. You can also attach fly shakers or shoo-flies to your horse's bridle. Fly shakers are like bangs on your horse's forehead; shoo-flies are horse hair tassles. They jiggle flies off your horse's face and head when he moves. There are also several types of fly masks that your horse can wear that will prevent flies from landing around his eyes. Also, your horse can wear a fly sheet, a cool, open-weave blanket that covers his body and prevents flies from getting at his skin.

Sanitation also includes keeping the mouse population in control in and around your horse's living quarters. Rodents such as mice and rats can cause damage and health problems if they are allowed to breed and become too numerous in your barn area. Rodents can carry bubonic plague and rabies. They can also chew expensive tack and destroy it. And if you let mice get started in your feed room, they will make a mess by chewing through your feed sacks.

All of your horse's feed should be stored in rodent-proof containers such as big garbage cans or bins. If you keep the grass around your barn trimmed and clean your barn regularly, you won't be providing good nesting sites for the mice. Don't use poison and bait to kill rodents because it is too dangerous if you have cats, dogs, or younger brothers and sisters. Having a few barn cats is much better than using poison. Cats are natural predators of mice, and just the presence of a few cats in the barn will keep mice away.

Emergency Health Care

If your horse gets sick or hurt, you should know enough so that you can talk to your veterinarian on the phone and tell him what you observe. If you make a daily health check, you will realize that something is wrong if your horse is "different" one day. For example, if your horse is usually standing by his feeder in the morning when you come out to feed, and one day you come out and he is lying down and doesn't get up, you know something is wrong.

Here are some of the most common problems that horses have. If your horse ever has these problems, you will be better able to care for him if you learn to recognize the symptoms.

Colic is similar to a stomach ache. Something has upset your horse's digestive system.

Symptoms: If your horse is suffering from colic, he will probably be either depressed or restless. A restless horse lies down and gets right back up again, or lies down and rolls over and over, or kicks at his belly with his hind legs, or turns his head around and looks at his sides. A depressed horse just stands or lies without moving, as though he has "given up" and just feels awful. He won't eat or drink at all. If your horse shows either of these types of symptoms, you'd better call your veterinarian and have the following information ready if he or she asks for it:

- When did you first notice this?
- Has the horse had a change in feed?
- Has he been drinking water?
- Did you see any fresh manure piles in his pen or stall?
- What is your horse's temperature, pulse, and respiration?

Treatment until the veterinarian arrives: After you have had a chance to talk with your veterinarian for a few minutes about your horse's condition, ask if you should walk your horse or not. In some cases of colic, it is helpful to walk the horse. The quiet exercise helps move whatever feed is causing the problem through the horse, and often you will see your horse poop or pass gas while you are walking him. When that happens, he will start feeling much better right away.

In other cases of colic, however, it is best *not* to walk the horse and instead to keep him very quiet. The better the information you provide your veterinarian, the better he or she will be able to tell you what to do until he or she gets there.

colic

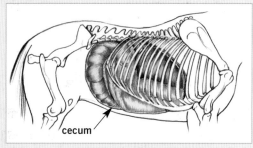

The right (off) side of the horse showing the cecum, where food can become impacted and cause colic.

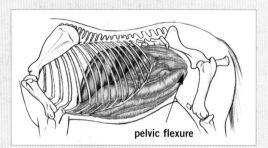

The left (near) side of the horse showing the pelvic flexure, an abrupt turn in the intestines which is another place where feed can get stuck and cause colic.

Heaves

Heaves describes a disease of the lungs usually found in horses over five years old. A horse with heaves has a hard time breathing. This condition might have originally been caused by an infection, but once the damage has been done to the lungs, the horse will always have the condition. Exercise, dusty pens, a dry climate, and moldy or dusty hay will make it much worse.

Symptoms: A horse with heaves coughs and has to use his belly muscles to help his lungs breathe.

Treatment: There is no cure for heaves, but a horse with a mild case of heaves can be made more comfortable if you wet his hay down before feeding it to him and keep him in a clean environment.

◯ Tying up (azoturia)

This is a condition in which a horse's body cramps, and it is painful for him to move. It usually occurs when a horse is fed too much grain and is not exercised regularly.

Symptoms: After a few days off, the horse is taken for a ride. Fifteen minutes later, his muscles lock up and he quits moving.

Treatment: If this happens, don't force the horse to move because it would probably tear his muscles. Send for help. Meanwhile, keep the horse quiet and warm. Put a blanket over his hindquarters. To prevent tying up, if you know you won't be riding your horse for a few days or more, decrease his grain ration.

◯ Lameness

Symptoms: Lameness is a problem that makes it difficult or painful for your horse to swing his legs forward or put weight on

lameness

Lame front leg: horse raises head to lift weight off the lame leg.

Lame hind leg: horse lowers head to shift the weight forward and off the lame leg.

his feet. Remember, however, that when horses rest, their favorite position is to stand on three legs. If you find your horse limping or holding up a leg, not wanting to put any weight on it, he might be lame.

Treatment: If your horse is truly sore on one of his feet or legs, here are the things you should check:

✔ First look for any obvious wounds on any of the legs. Some wounds are very sneaky, and unless you look very closely, you will miss them. Puncture wounds are like that. If your horse ran into a stick and it poked a hole in his leg, the wound would be very small, but it would be deep, and there might even be some pieces of the stick deep in his leg. So look carefully.

✅ Look at each hoof and pick it out, keeping your eyes open for rocks, wood splinters, nails, fence staples, wires, or anything else your horse could have picked up.

✅ While you are looking at each hoof, look at the shoes to see if they are loose or have shifted in position. One of the horseshoe nails might be causing a problem.

Most wounds that aren't going to be stitched can be cleaned with water from a hose. Hold the hose above the wound and let a gentle stream of water flow over the wound for two minutes.

⊃ Cut or other injury

Any time your horse gets seriously injured, call your veterinarian immediately. When your horse has a large or deep wound, the sooner your vet can treat and close it, the better chance your horse will have of healing. Some very small wounds will not require your veterinarian's care; in such a case you should ask an experienced horse person for advice on how to treat the wound.

If a wound is going to be stitched, don't do anything to the wound. Wait until the vet comes.

Treatment: The best way to clean most wounds is to let water from a hose run over the wound for a few minutes.

Never put a bandage on a wound until you have cleaned the wound thoroughly. And don't apply any home remedies, gook, or wound creams until you have had an experienced horseperson or veterinarian look at the wound.

⊃ Laminitis (founder)

When a horse first has this condition, it is called laminitis. After he has had it for a while, and it has affected his hooves, it is usually said, "The horse has foundered." It is a serious lameness of the hooves, which can happen when a horse eats too much grain or green pasture, when a hot horse drinks cold water, or when a horse is ridden too long or fast on a hard surface. Laminitis can also be caused by other circumstances, such as stress, foaling difficulties, or other trauma.

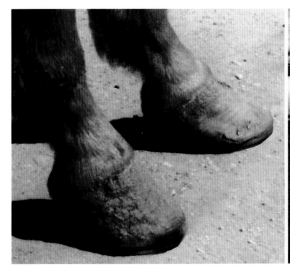

Hooves that are neglected and allowed to grow this long can be very painful for your horse and result in lameness.

Thrush looks like black tar, deep in the frog of your horse's hoof.

Symptoms: The horse's hooves get hot and painful, and the inner structures are damaged. The horse has difficulty moving and standing.

Treatment: If your horse has the symptoms of laminitis, call your veterinarian immediately. The vet will prescribe treatment. Many horses that have foundered once have a greater risk of foundering again and might always have lameness in their hooves.

❯ Thrush

Thrush is a disease of the frog area of the hoof. It is caused by wet, dirty stalls or pens. The filth lets a particular bacteria destroy the frog tissue.

Symptoms: You will know if your horse has thrush because when you pick out his hooves, it will smell awful! There will be a black gook something like tar deep in the clefts of his frog.

Treatment: If your horse has thrush, keep his hooves scrupulously clean and dry, and ask your farrier or veterinarian to treat the condition. They might recommend a solution for you to apply a few times to help clear it up faster.

❯ Navicular disease

Navicular disease is a painful condition of the front heels caused by poor hoof and leg conformation, poor shoeing, irregular hoof care, or being worked too hard or on poor footing.

Symptoms: The horse takes short, shuffling, stiff steps.

Treatment: Prevent navicular disease by finding a good horseshoer and keeping your horse on a regular shoeing schedule. If a horse already has navicular disease, good shoeing can help him be comfortable and useful.

chapter eight

Enjoying Your Horse

You can enjoy your horse or pony in many ways with your family and friends or with a *youth* group. First, master the care of your horse. Then devote much time to learning how to ride properly. Only after you have become a well-rounded horseman or horsewoman should you consider showing.

The purpose of showing is to allow a good rider to demonstrate the thoroughness of horse care and training abilities. You should be experienced and your horse should be well trained before you enter the show ring. Then you will have a safe and enjoyable experience. Showing requires a lot of at-home preparation — not just a quick bath the day before the show but careful planning and training for months ahead of time.

Some youth groups are designed to teach you about horse care and horsemanship as well as competition skills. Remember, when you participate in a group activity, don't be a show-off with your horse. That usually leads to someone getting hurt.

If you are entered in a show, no matter how small or large, do your best and hope to win, but if you lose, be sure to grin. Congratulate the winner and think about what you need to work on to do a better job next time. Put fun before winning. That's good *sportsmanship*.

If you have family members or friends that are also interested in horses, you can find many ways to enjoy your horses together. Pleasure trail riding is one of the most satisfying and relaxing ways to use a horse. It can also be exciting, depending on where you go to ride. It's great fun to plan a day of trail riding with some friends and meet at a park or trail area. If you can't get to a park or trail easily, you can plan get-togethers riding through pastures, along safe roadways, and in various arenas.

A 4-H Club meeting is a great opportunity to learn about every aspect of horsekeeping.

Local Horse Groups

Many activities are best enjoyed with an organized group. Look for flyers of local horse groups in your tack and feed stores or see the suggestions given on the following page. Call up the group's president or secretary and find out when and where the next meeting will be held. Attend the meeting as a visitor, and see if the group has the same horse interests you do. Some groups focus on trail riding, others on education and clinics, and others on horse shows.

Participating in 4-H. Many states and provinces run a program called 4-H, which has been in existence since the early 1900s. The 4-H Club covers all sorts of agriculture and livestock interests; the horse project is just one of them.

Usually, a 4-H group is made up of from three to twenty members and one or two volunteer leaders, who are usually parents and who may or may not have horse experience. The group picks a club name and usually meets once a month in the winter and more often in the spring, summer, and fall.

Since many of the programs will probably involve riding and using your horse, be aware that you will need to trailer your horse to the meeting site.

Many horse activities, such as trail riding, are best enjoyed with a group or a good friend!

The 4-H Advancement Levels. This is a systematic program that helps you progress from beginner to more experienced horseperson. Everyone must begin with Level 1, the novice or beginner level, and by taking written tests and demonstrating riding and other required skills, you can pass to the next level.

The 4-H Horse Show. Usually every county and state has at least one 4-H show each year, and some can have as many as twenty classes! Some of the more common classes are Showmanship, Western Horsemanship, and English Equitation. In Showmanship, you will be judged on your ability to exhibit your horse to a judge at halter. In Western Horsemanship, you ride your horse with Western tack and attire to demonstrate your riding skill. English Equitation requires that you use English tack and attire to demonstrate your riding skills.

Your local 4-H show might also have halter classes, gymkhana (barrel racing, pole bending, and so on), Western pleasure, hunter under saddle, and other classes.

Participating in Pony Club. The

United States Pony Club (USPC) was established in 1954, and today there are over 500 clubs and 12,000 members, with 90 percent of the members being girls.

Don't be fooled by the word "pony" — this doesn't mean that all members own ponies; in fact, most members actually own horses. The USPC was designed to teach English riding, mounted sports, and

Participation in Pony Club includes learning safety during mounted sports.

the care of horses and ponies to those under twenty-one years of age, and it also provides opportunities for competition in various horse sports.

The USPC's mission is to provide a program that develops responsibility, moral judgment, leadership, and self-confidence, while stressing safety and common sense in clothing and equipment.

USPC competitions include a variety of local, regional, and national competitions. Members can compete as individuals and as members of a team against USPC standards of proficiency. USPC has four ratings from A to D with different levels within the ratings. As you progress from one level to another, you establish a good, solid base of knowledge and experience, from horse care to riding and training skills to teaching skills.

To attend horse activities, your horse will need to become accustomed to traveling in a horse trailer.

Participating in FFA. The Future Farmers of America (FFA), organized in 1928, has more than 400,000 members who participate in programs related to agriculture. The FFA is dedicated to helping young people develop their potential for premier leadership, personal growth, and career success through agriculture education.

The FFA Agricultural Proficiency Award Program helps members develop their interest in a particular agricultural subject, and the Horse Proficiency Program gives members an idea of what is involved in breeding and raising horses. It can also include experience programs that may result in career opportunities in the horse business. You must be in at least tenth grade to become an FFA member.

Horse Shows

Horse shows are competitive events sponsored by a breed or performance association or local horse group. There are many types of shows and a wide variety of classes in horse shows.

Schooling show. This is a "practice" show that is usually held before the regular show season begins. Sometimes you can dress more casually than you do for a formal horse show, and often you can use training tack. And if it's in the winter, it's okay if your horse is fuzzy with his winter coat.

Breed show. This is only for registered horses of a certain breed. For example, an American Quarter Horse

4-H and FFA projects encourage you to use your horse for career-oriented activities.

At a schooling show, you can dress and groom more casually than for a formal show.

Association show is only for horses that are registered with the AQHA.

Association Show. This is a multi-breed show sponsored by a horse organization. For example, a National Reining Horse Association (NRHA) show is devoted solely to reining and is open to horses of any breed.

Youth show. This is a show for riders eighteen years and under. It is usually a part of a breed show or an association show except on the regional or national level, when the youth championships are held separately.

Open show. This type of show is open to all ages and breeds of horses and all ages of riders. It is usually sponsored by a local club or riding group.

get connected

For more information on your local 4-H Horse Program, contact your county extension office. In the business listing or government listing section (not the yellow pages) of your phone book, look under the name of your county. It will say something like:

Larimer County of
Cooperative Extension Office
Agriculture, 4-H/Youth 555-7400

To find the name and location of the horse groups nearest you and for more information, contact:

The United States Pony Clubs, Inc.
The Kentucky Horse Park
4041 Iron Works Parkway
Lexington, KY 40511-7669
www.ponyclub.org
(859) 254-7669

National FFA Center
P.O. Box 68960
Indianapolis, IN 46268-0960
www.ffa.org
(317) 802-6060

USA Equestrian
4047 Iron Works Parkway
Lexington, KY 40511-8483
www.usaequestrian.org
(859) 258-2472

American Quarter Horse Youth Assoc.
P.O. Box 200
Amarillo, TX 79168
www.aqha.org.
(806) 376-4811

horse associations

There are many other breed and performance associations with youth programs. If you write to the **American Horse Council** you can purchase a Horse Industry Directory. This big 175-page book lists the names and addresses of all horse organizations in the United States:

> **American Horse Council**
> 1700 K Street, NW, Suite 300
> Washington, DC 20006
> (202) 296-4031
> www.horsecouncil.org/ahc.html

In the Horse Industry Directory, you'll find the addresses of the following associations that have youth classes or programs:

- American Morgan Horse Association
- American Paint Horse Association
- Appaloosa Horse Club
- International Arabian Horse Association
- International Buckskin Horse Association
- National Cutting Horse Association
- National Reining Horse Association
- North American Trail Ride Conference
- Palomino Horse Breeders Association
- Pony of the Americas Club
- United States Combined Training Association
- United States Dressage Federation

Horse Show Organizations

USA Equestrian, formerly known as the American Horse Shows Association, was founded in 1917 and serves to regulate competitions for many breed and discipline groups. A discipline is a particular style of horse use, such as dressage, jumping, driving, Western riding, or hunter classes. Most of the members of USA Equestrian participate in hunter/jumper classes, dressage, or three-day eventing. Junior riders (those under the age of eighteen) can participate in many of the events plus equitation classes, pony classes, and vaulting.

Many breed associations have special Youth Classes in Showmanship and English and Western riding.

Pony Club competitions emphasize good sportsmanship, good horsemanship, and safety.

The American Quarter Horse Association (AQHA) was established in 1940 to record and preserve pedigrees of registered American Quarter Horses and to encourage members to participate in various activities with their Quarter Horses. There are more than 330,000 members and more than 2,700 AQHA shows every year worldwide.

The American Quarter Horse Youth Association (AQHYA) is designed for members under eighteen years of age. Usually an AQHYA show is held in conjunction with a regular AQHA show, with twenty-three youth-only events, including Western, English, showmanship, and halter classes. If you are new to showing, you can compete in special classes for novices.

There is usually at least one AQHYA youth club in each state and province. The purpose of the local clubs is for members to meet people, gain confidence, and learn how AQHA works. Members can participate in public speaking contests, horse judging contests, and marketing contests and can apply for scholarships and awards.

The AQHA also has a Horseback Riding Program for members who enjoy trail riding. After you enroll, you keep a log of the number of hours spent riding your American Quarter Horse, and when you accumulate 50 hours, you receive a special award. You can earn additional awards for 100, 250, 500, 750, 1,000, 2,000, and 3,000, 4,000, and 5,000 hours as well.

Appendix

Horse Health Record

Here's a form you can use to keep track of your horse's health. Make a photocopy of this page for each horse. Cut out and tape the two sides to the front and back of an index card, or tape them to the front of your horse's stall.

HORSE HEALTH RECORD

Name of horse_____

Date of birth_____ Height _____ Weight _____

Breed_____

Color and markings_____

_____

Owner_____ Phone_____

Veterinarian _____ Phone_____

Farrier_____ Phone _____

Feeding instructions:

---

front of index card

Deworming		Farrier		Vaccinations	
date	product used	date	service	date	vaccine

Date of last coggins _____ Date of health certificate_____

back of index card

Horse Care Calendar

winter/spring

Early Winter

✔ Deworm for all worms.

Mid-Winter

✔ Be sure your horse is drinking plenty of water during the cold weather.

✔ Monitor his weight (see Late Summer).

Late Winter

✔ Deworm for all worms and bots.

Early Spring

✔ Spring dental check up.

✔ This is usually a wet period so be sure your horse's hooves stay clean and dry.

✔ Take horses off winter pasture so they don't tear up the soft earth with their hooves.

✔ Let the pasture grasses grow.

tip:

Remember, every day your horse should have an inspection that includes his hooves.

Horse Care Calendar

spring

Mid-Spring

✅ Check for ticks

✅ Deworm for all worms and bots.

✅ Vaccinate at least for sleeping sickness, tetanus, influenza, and rhinopneumonitis, and maybe for strangles, rabies, and other diseases (ask your vet).

✅ Mares start coming into heat.

✅ Begin a spring conditioning program.

✅ Gradually start feeding your horse some grain according to the amount of work he will be doing.

✅ Keep an eye on pastures for the appearance of poisonous weeds.

✅ Shedding begins.

Late Spring

✅ Check your fences and gates and make repairs.

✅ Begin introducing your horse gradually to spring pasture.

✅ Be sure your horse has free-choice salt and minerals in stalls, pens, and pastures.

✅ Heavy shedding of winter coat and growth of new short, summer coat occurs.

Horse Care Calendar

summer

Early Summer

✓ Deworm for all worms; double-dose for tapeworms.

✓ Rotate pastures.

✓ Provide shelter from the summer sun.

Mid-summer

✓ Rotate pastures.

✓ Consider using a fly sheet and fly mask.

✓ Check your horse's weight to be sure he is not getting too fat from the summer pasture.

✓ Buy your year's supply of hay.

Late Summer

✓ Deworm for all worms.

✓ Rotate pastures.

✓ Shedding of the summer coat begins, and the long winter coat starts growing.

✓ Use a weight tape and feel your horse's ribs once a month from now until spring. You won't be able to tell if your horse is getting too thin by just looking at him because his fuzzy winter coat might make him look fatter than he really is.

tip:
Every 6–8 weeks your horse should have professional farrier care, either trimming or shoeing.

Horse Care Calendar

fall

Early Fall

✅ Rotate pastures.

✅ Remove bot eggs.

✅ Perhaps begin blanketing at night.

✅ Schedule a fall dental check-up and maybe rhino and influenza booster shots.

✅ It you live where there is winter cold, allow your horse to gain 5 percent but no more than 10 percent of his body weight. A 1200-pound adult horse could gain 60–120 pounds in the early fall. This extra flesh and fat will provide added insulation and an energy and heat reserve when weather is particularly bad.

✅ Take horses off winter pasture before they have eaten it down.

✅ Check your fences and make repairs.

Mid-Fall

✅ Deworm for worms and bots after the first hard frost.

✅ Remove bot eggs.

✅ Mares stop coming into heat.

✅ You might need your horseshoer to put winter shoes on your horse.

Late Fall

✅ For every 10 degrees Fahrenheit below freezing (32°F), increase hay portion of ration by 10 percent.

✅ Decrease the horse's grain if his activity level is low.

✅ Be sure he has good shelter from the snow, wet, cold, and wind.

✅ You can start turning your horse back on pasture once there is a blanket of snow covering it.

GLOSSARY

Action (n.). The degree of flexion of the joints of the legs during movement; also reflected in head, neck, and tail carriage. High, snappy action is desired in some classes while easy, ground-covering action is the goal in other classes.

Age (of the horse) (n.). Computed from January 1 of the year in which the horse is foaled.

Aid (n.). An aid is an action by a part of the rider's or handler's body to a part of the horse's body to cause the horse to react in a particular way. An aid is almost never used alone but always used in conjunction with other aids. A rider's natural aids are his or her mind, seat, weight, upper body, legs, hands, and voice. The combined use of all of the rider's aids simultaneously produce a smooth, balanced response from a horse. A handler's natural aids are the mind, hands, and overall body language. Examples of Artificial Aids (which are extensions, reinforcements, or substitutions for the natural aids) are whips, spurs, and nosebands.

Appointments (n.). Tack and equipment (attire is sometimes included).

Attire (n.). The rider's clothes.

Back (n. and v.). A two-beat diagonal gait in reverse.

Bad habit (n.). Undesirable behavior during training or handling. Examples are rearing, halter pulling, striking.

Balance (n.). In regard to conformation, desirable proportions.

Balk (v.). To refuse or cease to move forward.

Barn sour (adj.). Herd-bound; a bad habit that may result in a horse bolting back to the barn or to his herd-mates.

Bay (n. and adj.). A body color ranging from tan to reddish-brown, with black mane and tail, and usually black on the lower legs.

Beat (n.). A single step in a gait, involving one leg or two. For example, the walk is a 4-beat gait, with each beat stepped off by a single leg, one at a time, 1-2-3-4. The trot is a 2-beat gait, stepped off by two legs landing at the same time, 1-2.

Biting (n.). A bad habit common to young horses, stallions, and spoiled horses. It can result from hand-fed treats, petting, or improper training.

Black (adj. or n.). A body color that is true black over the entire body, but may have white leg and face markings.

Blemish (n.). A visible defect that does not affect serviceability.

Bloodlines (n.). The family lineage.

Blue roan (n. and adj.). A body color that has a uniform mixture of black and white hairs all over the body.

Bolt (v.). Gulp feed without chewing; run away with rider.

Bot block (n.). A rough, porous "stone" used to scrub off bot eggs.

Bot fly (n.). A fly that looks like a bee and lays eggs in a horse's hair.

Bowed tendon (n.). The damage to a tendon usually caused by overstretching due to improper conditioning, overwork, or an accident.

Breed character (n.). The quality of conforming to the description of a particular breed.

Breed registry (n.). An organization that keeps track of all the ancestors and current members of a breed.

Broodmare (n.). A mare used for breeding.

Brown (adj. and n.). A body color with mixed brown and black hair, with black mane, tail and legs.

Buckskin (adj. and n.). A body color that is tan, yellow, or gold with black mane, tail, and lower legs.

Canter (n. and v.). The English term for a three-beat gait with right and left leads. The canter has the same foot fall pattern as the lope.

Chestnut (n. and adj.). A color in which the body, mane, and tail are various shades of brown; oval, horny growth on the inside of the legs (see Parts of the Horse on pages 8–9).

Cob (n.). A small horse.

Coggins test (n.). A laboratory blood test used to detect previous exposure to equine infectious anemia.

Cold-blooded (adj.). Refers to horses having ancestors that trace to heavy war horses and draft breeds. Characteristics might include more substance of bone, thick skin, heavy hair coat, shaggy fetlocks, and blood type that makes it suitable for slow, hard work.

Colic (n.). Intestinal discomfort, which can rage from a mild stomachache to a life-threatening violent frenzy.

Color (n.). Description of body coat color and pattern.

Colt (n.). A young male horse to age four.

Conditioning (n.). The art and science of preparing a horse mentally and physically for a particular use such a pleasure riding, competitive trail riding, or showing.

Conformation (n.). The physical structure of a horse, which is compared to a standard of perfection or an ideal.

Cribbing (n. and v.). A vice whereby a horse anchors its teeth onto an object, arches its neck, pulls backward, and swallows air. It can cause the horse to lose weight, suffer tooth damage, and other physical disturbances. It can be a contagious habit.

Crossbred (n. and adj.). A horse that has one parent of one breed and the other parent of another breed.

Cross-tie (n. and v.). A means of tying a horse in which a chain or rope from each side of an aisle is attached to the side rings of the horse's halter.

Cryptorchidism (n.). The retention of one or both testicles in the abdominal cavity.

Cue (n.). A single signal, often made up of several aids, from the rider or handler that tells a horse what to do. Often used in performing tricks.

Dam (n.). Mother of a horse.

Diagonal (n.). A pair of legs at the trot, such as the right front and the left hind. When posting, the rider sits as the inside hind hits the ground or "rise and fall with the (front) leg on the wall." Riding across the diagonal is a maneuver from one corner of an arena to another through the center.

Disunited (adv.). Cantering or loping on different leads front and hind.

Dock (n.). The flesh and bone part of the tail.

Draft horse (n.). A horse of one of the breeds of "heavy horses" developed for farm or freight work, such as Percheron, Belgian, and Clydesdale. Draft horses weigh 1,500–2,200 pounds and can be as tall as 17 hands. They are generally not suitable for riding.

Driving (adj. and v.). Description of an event where a horse or pony pulls a wagon or cart.

Dun (n. and adj.). A yellow or gold body and leg color, often with a black or brown mane and tail, usually with a dorsal stripe and stripes on legs and withers.

English (adj.). Referring to riding with English tack and attire.

Equestrian (adj.). Of or pertaining to horsemen or horsemanship; a rider

Equitation (n.). The art of riding.

Farrier (n.). A person who shoes horses.

Fetlock (n.). The joint between a horse's pastern and cannon.

Filly (n.). A female horse to age 4.

Flank (n.). The area of a horse's barrel between the rib cage and the hindquarters.

Flaxen (adj.). A golden mane or tail on a darker-bodied horse.

Floating (n. and v.). The process of filing off sharp edges of a horse's teeth.

Foal (n.) A male or female horse or pony under 1 year of age.

Forehand (n.). That portion of the horse from the heart girth forward.

Forelock (n.). The hair growing between a horse's ears that falls on the forehead; a horse's "bangs."

Founder (n. and v.). Another word for laminitis, a serious disease affecting a horse's hooves and often caused by a horse's eating too much grain or green pasture.

Frog (n.). The thick, triangle-shaped tissue on the bottom of a horse's hooves.

Gait (n.). A specific pattern of foot movements such as the walk, trot, and canter.

Gaited horse (n.). An animated horse such as the Arabian, American Saddlebred, Morgan, or Tennessee Walking Horse with flashy gaits.

Gelding (n.). A male horse that has been castrated (had its testicles removed).

Grade (adj. and n.). An unregistered horse.

Gray (adj. and n.). A color in which the skin is black, and the hair is a mixture of black and white.

Green (adj.). An inexperienced horse or rider, relatively speaking.

Ground training (n.). When the trainer works the horse from the ground, rather than being mounted. Includes in-hand work, barn manners, longeing, and ground driving.

Grullo (adj. and n.). A type of dun with a smoky or mouse-colored body, and usually having a black mane, tail, lower legs, and dorsal stripe.

Halter class (n.). Conformation class.

Halter pulling (n.). A bad habit in which a horse pulls violently backward on the halter rope when tied.

Hand (n). Horses are measured from the highest point of the withers to the ground in units called hands. One hand equals four inches. 14.2 means (14 hands x 4 inches) + 2 inches, which is 56 inches + 2 inches = 58 inches.

Haunches (n.). Hindquarters.

Head shy (adj.). Description of a horse who shies away from having his head touched.

Heart girth (n.). The measurement taken around the horse's barrel just behind the front legs.

Heat (n.). The part of a mare's reproductive cycle when she is ready to mate with a stallion.

Heaves (n.). Damage to the lungs, resulting in labored breathing.

Herd-bound (adj.). When a horse is too dependent on being with other horses and doesn't want to be separated from them.

Honest (adj.). A quality in a horse which makes him dependable and predictable.

Horse (n.). An equine over 14.2 hands.

Horsemanship (n.). Exhibition of a rider's skill, usually referring to the Western style of riding.

Hot-blooded (adj.). Refers to horses having ancestors that trace to Thoroughbreds or Arabians. Characteristics might include fineness of bone, thin skin, fine hair coat, absence of long fetlock hairs, and blood type that makes it well-suited for speed and distance work.

Hunter (n.). A type of horse, not a breed, which is suitable for field hunting or show hunting.

In-hand class (n.). A class in which the horse is led by the exhibitor.

Jog (n. and v.). A slow Western trot.

Junior (n.). A rider under eighteen years of age as of January 1.

Larvae (n.). Insects or parasites that have hatched from eggs but are not yet mature.

Lead (n.). A specific footfall pattern at the canter or lope in which the inside legs of the circle reach farther forward than the outside legs. When working to the right on the right lead, the horse's right foreleg and right hind leg reach farther forward than the left legs. If a horse is loping in a circle to the right on the left lead, he is either on the wrong lead or he is counter-cantering.

Liver chestnut (n. and adj.). A very dark red chestnut color, with mane, tail, and legs the same color as the body or flaxen.

Longe (v.). To work a horse in a circle usually on a 30-foot line around you at various gaits.

Lope (n. and v.). A three-beat gait: (1) an initiating hind leg; (2) a diagonal pair including the leading hind leg and the diagonal foreleg; and (3) the leading foreleg.

Manners (n.). The attitude and habits of a horse.

Mare (n.). A female horse over age four.

Markings (n.). White on the face or legs of a horse.

Muzzle (n.). The end of a horse's face, including the nose, nostrils, and lips.

Near side (n.). The horse's left side.

Novice (n.). In general, an inexperienced horseperson.

Off side (n.). The horse's right side.

Pacing (n.). Continuous stall or pen walking, often an unhappy horse's reaction to confinement. Also a two-beat lateral gait of standard bred race horses.

Paddock (n.). A small pasture.

Paint (n. and adj.). A breed of horse with large blocks of white and black or white and brown; paint is a coat pattern on any breed of horse that is similar to that on a Paint horse.

Palomino (n. and adj.). A breed of horse that has a golden body color and a light to white mane and tail; a palomino is a horse with coloring similar to that of a Palomino horse.

Panic snap (n.). A safety snap often used in horse trailers and cross-ties. The design allows the snap to be released even if there is great pressure on it.

Parasite (n.). A harmful organism that lives in or on another organism.

Park horse (n.). A horse with a brilliant performance, style, presence, finish, balance and cadence, and usually animated gaits.

Parrot mouth (n.). An unsoundness of the teeth that is characterized by an extreme overbite.

Pattern (n.). A prescribed order of maneuvers in a particular class such as horsemanship, equitation, reining, or trail.

Pawing (n.). A bad habit usually caused by nervousness and/or improper ground training; can also be a sign of colic.

Pecking order (n.). Social rank of each horse in a group; one horse is the boss and the others find their place.

Pedigree (n.). A listing of a horse's ancestors.

Pen (n.). An outdoor, non-grassy living space that is approximately 24 feet long and 24 feet wide or larger.

Points (n.). The coloring of the legs, mane, and tail.

Poll (n.). The junction of the vertebrae with the skull located between a horse's ears; an area of great sensitivity and flexion.

Pony (n.). A horse that stands 58 inches (14.2 hands) or less.

Posting (n.). A way to ride the English trot; See *diagonal.*

Presence (n.). Personality, charisma.

Pulse (n.). Heart rate. Normal adult resting heart rate varies among horses but is usually 40 beats per minute.

Pupae (n.). The stage of development between the bot egg and the bot fly.

Purebred (n. and adj.). A horse of pure ancestors of a particular breed.

Quality (n.). Overall degree of merit: flat bone and clean joints, refined features and fine skin and hair coat.

Rearing (n.). A bad habit in a horse, of raising up on his hind legs when he is being led or ridden. An extremely dangerous habit that should be dealt with by a professional only.

Red roan (n. and adj.). A mixture of red and white hairs all over a horse's body, with red, black, or flaxen mane and tail. Also called *strawberry roan.*

Refinement (n.). Quality appearance, indicating good breeding.

Registered (adj.). A horse of purebred parents that have numbered certificates with a particular breed organization.

Rein-back (n.). To back up; a two-beat diagonal gait in reverse.

Respiration rate (n.). Number of breaths per minute. Normal adult respiration rate varies among horses but is usually 12 to 15 breaths per minute. One breath consists of an inhalation and an exhalation.

Ringbone (n.). An arthritic unsoundness of the pastern joint.

Ring sour (adj.). A poor attitude in a horse who does not enjoy working in an arena and looks for ways to leave the arena or quit working.

Roached (adj.). A mane or tail that has been clipped to the skin.

Roan (n. and adj.). A horse color resulting from a mixture of white and black or white and red hairs all over the body.

Roaring (n.). A breathing disorder.

Run (n.). A long, narrow fenced-in area usually attached to a stall.

Sand colic (n.). A digestive disorder that occurs when a horse eats sand or dirt with his feed.

Sheath (n.). The skin folds that encase a horse's penis.

Showmanship (n.). An in-hand class that is judged on the exhibitor's ability to show his horse.

Shying (n.). A horse spooking or becoming startled by a movement or object. It may or may not include a sudden jump sideways, or bolting.

Sire (n.). Father of a horse.

Sorrel (n.). A reddish or coppered body with mane and tail the same color as the body.

Sound (adj.). Having no defect, visible or unseen, that affects serviceability; the state of being able to perform without hindrance.

Spavin (n.). An unsoundness of the hock which can involve soft tissues (bog spavin) or bone (bone spavin or jack spavin).

Spayed mare (n.). A neutered female horse.

Splint boots (n.). Protective covering worn around the cannons of the front legs to prevent injury.

Splints (n.). A term referring to bony enlargements at various points along either of the splint bones, located on each side of the cannon bone.

Spooky (adj.). An easily startled horse.

Sport horse (n.). A purebred or crossbred horse suitable for dressage, jumping, eventing, or endurance.

Stallion (n.). A male horse (not gelded).

Step (n.). A beat.

Stock horse (n.). A Western-style horse of the Quarter Horse type.

Strawberry roan (n. and adj.) See *red roan.*

Stride (n.). The distance traveled in a particular gait, measured from the spot where one hoof hits the ground to where it next lands. Ten to twelve feet is the normal length of stride at a canter, for example.

Striking (n.). A bad habit of reaching out with a front foot so as to hit the handler, equipment, or another horse. A problem calling for professional help.

Stud (n.). A stallion used for breeding.

Substance (n.). Strength and density of bone, muscle, and tendons or an indication of large body size.

Suckling (n.). A foal that is still with its mother; it has not been weaned; usually it is under four months of age.

Suitability (n.). Appropriateness for a particular purpose and/or a type or size of rider.

Sullen (adj.). Sulky, resentful, withdrawn.

Tack (n.). Horse equipment or gear.

Tail rubbing (n.). A habit that may originate from anal or skin itch or a dirty sheath or udder. Even when the cause is removed, the habit often persists.

Temperament (n.). The general consistency with which a horse behaves.

Temperature (n.). Normal adult temperature varies among horses, but will usually range in degrees from 99.5°F to 100.5°F.

Thoroughbred (n.). The breed of horse registered with the Jockey Club. Not meant to be used as a synonym for purebred.

Thrush (n.). A disease of the hoof often associated with unsanitary conditions, which causes decomposition of the frog and other hoof structures.

Topline (n.). The proportion and curvature of the outline of a horse's neck, back, and croup; a line from poll to tail-head.

Tractable (adj.). A quality in a horse's disposition that makes him cooperative and trainable.

Travel (n.). The path of the flight of each limb during movement.

Trot (n. and v.). A two-beat diagonal gait.

Twitch (n.). A means of restraint. A nose twitch is often a wooden handle with a loop of chain, applied to the horse's upper lip.

Tying up (n.). A form of metabolic muscle stiffness caused from irregularity in feed and work schedules.

Type (n.). A particular style of horse with certain characteristics that contribute to its value and efficiency for a particular use.

Udder (n.). The mammary glands or teats of a female horse.

Underline (n.). The length and shape of the line from the elbow to the sheath or udder.

Unsoundness (n.). A defect that may or may not be seen but does not affect serviceability.

Vice (n.). Abnormal behavior in the stable environment that results from confinement or improper management and can affect a horse's usefulness, dependability, and health. Examples are cribbing and weaving.

Walk (n. and v.). A four-beat flat-footed gait.

Weanling (n.). A foal that has been separated from its mother; usually four to twelve months of age.

Weaving (n.). Rhythmic swaying of weight from one front foot to the other when confined. Can be socially contagious.

Western (adj.). Referring to riding with Western tack and attire.

Withers (n.). The part of the horse's spine where the neck joins the back.

Wood chewing (n.). A common vice that damages facilities and can cause abnormal wear of teeth and possible complications from wood splinters.

Yearling (n.). A male or female horse or pony that is one year old.

Youth (n. and adj.) An exhibitor eighteen years of age and under. Additional age divisions are often created to separate children further.

RECOMMENDED READING

Damerow, Gail. *Fences for Pasture & Garden* (Storey Books, 1992)

Haas, Jessie. *Safe Horse, Safe Rider: A Young Rider's Guide to Responsible Horsekeeping* (Storey Books, 1994)

Harris, Susan. *Grooming to Win* (Hungry Minds, Inc., 1991)

Harris, Susan. *The United States Pony Club Manual of Horsemanship* (Hungry Minds, Inc., 1994)

Hill, Cherry. *Becoming an Effective Rider: Developing Your Mind and Body for Balance and Unity* (Storey Books, 1991)

Hill, Cherry. *From the Center of the Ring* (Storey Books, 1988)

Hill, Cherry. *Horse for Sale: How to Buy a Horse or Sell the One You Have* (Hungry Minds, Inc., 1995)

Hill, Cherry. *Horsekeeping on a Small Acreage: Facilities Design and Management* (Storey Books, 1990)

Hill, Cherry. *Horse Handling and Grooming: A Step-by-Step Photographic Guide to Mastering Over 100 Horsekeeping Skills* (Storey Books, 1997)

Hill, Cherry. *Horse Health Care: A Step-by-Step Photographic Guide to Mastering Over 100 Horsekeeping Skills* (Storey Books, 1997)

Hill, Cherry. *Stablekeeping: A Visual Guide to Safe and Healthy Horsekeeping* (Storey Books, 2000)

Hill, Cherry. *Trailering Your Horse: A Visual Guide to Safe Training and Traveling* (Storey Books, 2000)

Hill, Cherry. *101 Arena Exercises: A Ringside Guide for Horse & Rider* (Storey Books, 1995)

Cherry Hill and Richard Klimesh, *Maximum Hoof Power: How to Improve Your Horse's Performance through Proper Hoof Management* (Trafalgar Square, 1999)

Lon D. Lewis, *Feeding and Care of the Horse* (Lippincott Williams & Wilkins, 1995)

INDEX

Page numbers in **boldface** indicate a chart;
Page numbers in *italics* indicate a photograph or illustration.